Green Chemistry

ACS SYMPOSIUM SERIES 626

Green Chemistry

Designing Chemistry for the Environment

Paul T. Anastas, EDITOR

Tracy C. Williamson, EDITOR

Office of Pollution Prevention and Toxics
U.S. Environmental Protection Agency

Developed from a symposium sponsored
by the Division of Environmental Chemistry, Inc.,
at the 208th National Meeting
of the American Chemical Society,
Washington, DC,
August 21–25, 1994

American Chemical Society, Washington, DC 1996

Library of Congress Cataloging-in-Publication Data

Green chemistry: designing chemistry for the environment / Paul T. Anastas, editor, Tracy C. Williamson, editor.

p. cm.—(ACS symposium series, ISSN 0097–6156; 626)

"Developed from a symposium sponsored by the Division of Environmental Chemistry, Inc., at the 208th National Meeting of the American Chemical Society, Washington, D.C., August 21–25, 1994."

Includes bibliographical references and indexes.

ISBN 0–8412–3399–3

1. Environmental chemistry—Industrial applications—Congresses. 2. Environmental management—Congresses.

I. Anastas, Paul T., 1962– . II. Williamson, Tracy C., 1963– III. American Chemical Society. Division of Environmental Chemistry, Inc. IV. American Chemical Society. Meeting (208th: 1994: Washington, D.C.) V. Series.

TP155.G635 1996
660′.281—dc20 96–162
CIP

Second printing 1998

This book is printed on acid-free, recycled paper.

PRINTED IN THE UNITED STATES OF AMERICA

1995 Advisory Board

ACS Symposium Series

Foreword

THE ACS SYMPOSIUM SERIES was first published in 1974 to provide a mechanism for publishing symposia quickly in book form. The purpose of this series is to publish comprehensive books developed from symposia, which are usually "snapshots in time" of the current research being done on a topic, plus some review material on the topic. For this reason, it is necessary that the papers be published as quickly as possible.

Before a symposium-based book is put under contract, the proposed table of contents is reviewed for appropriateness to the topic and for comprehensiveness of the collection. Some papers are excluded at this point, and others are added to round out the scope of the volume. In addition, a draft of each paper is peer-reviewed prior to final acceptance or rejection. This anonymous review process is supervised by the organizer(s) of the symposium, who become the editor(s) of the book. The authors then revise their papers according to the recommendations of both the reviewers and the editors, prepare camera-ready copy, and submit the final papers to the editors, who check that all necessary revisions have been made.

As a rule, only original research papers and original review papers are included in the volumes. Verbatim reproductions of previously published papers are not accepted.

Contents

Educational Tools

Indexes

Preface

GREEN CHEMISTRY focuses on the design, manufacture, and use of chemicals and chemical processes that have little or no pollution potential or environmental risk and are both economically and technologically feasible. The principles of green chemistry can be applied to all areas of chemistry including synthesis, catalysis, reaction conditions, separations, analysis, and monitoring.

The chemical industry in the United States releases more than 3 billion tons of chemical waste each year to the environment. Industry then spends $150 billion per year in waste treatment, control, and disposal costs. The challenge for chemists involved at all stages of chemical design, manufacture, and use is to make incremental changes that, when summed, will achieve significant accomplishments in the design of new products and processes that are less polluting and hazardous to the environment.

The symposium upon which this book is based was organized by Joseph J. Breen and Allan Ford under the auspices of the Division of Environmental Chemistry, Inc. This book is composed primarily of topics that were presented at sessions of the symposium that were chaired by the editors of this volume. In addition, presentations from another session of the same symposium that focused on environmentally benign chemistry research in the international arena, chaired by Steven Hassur, have also been included.

This book presents the current research efforts and recent results of leaders in the field of green chemical syntheses and processes. The projects described cover a range of topics that are broadly applicable to the chemical industry as well as to chemical education. As such, this book should appeal to chemists from academia, industry, and government who are involved in fundamental research, methods development and application, education, and decision making. Our hope is that this book will provide a wealth of information to chemists involved in chemical synthesis and processing at the research, applied, and management levels and will also act as a catalyst in stimulating many more chemists to become involved in the design and use of chemical syntheses and processes in an environmentally responsible manner.

Disclaimer

We edited this book in our private capacities. No official support or endorsement of the U.S. Environmental Protection Agency is intended or should be inferred.

Acknowledgments

We thank the many people who contributed their time and efforts toward making this volume possible. The dedication of Joseph Breen in furthering the cause of green chemistry through all avenues and specifically for his role in organizing the Design for the Environment Symposium is valued and appreciated. We also recognize the Division of Environmental Chemistry, Inc., and Allan Ford for their contributions to the symposium. The assistance of Margaret Cavanaugh and Maria Burka in identifying individuals for the original symposium sessions is much appreciated. We also thank Steven Hassur for his role in organizing the international session of the symposium.

Chemists who dedicated their time to provide insight and support for this book include: Steven DeVito, Russell Farris, Carol Farris, Daniel Lin, Daniel Bushman, Jenny Tou, Caroline Weeks, Diana Darling, Steven Hassur, Paul Tobin, Paul Bickart, Gregory Fritz, and Fred Metz. Many thanks to Rhonda Bitterli, whose guidance was essential to getting this volume initiated. Thanks to the ACS Books Department Staff, including Barbara Pralle and Charlotte McNaughton. And most of all, thanks to each of the authors whose outstanding efforts have made this volume so valuable.

PAUL T. ANASTAS
TRACY C. WILLIAMSON
Office of Pollution Prevention and Toxics
U.S. Environmental Protection Agency
Washington, DC 20460

December 12, 1995

Chapter 1

Green Chemistry: An Overview

Paul T. Anastas and Tracy C. Williamson

Office of Pollution Prevention and Toxics, U.S. Environmental Protection Agency, Mail Code 7406, 401 M Street, S.W., Washington, DC 20460

Green Chemistry is an approach to the synthesis, processing and use of chemicals that reduces risks to humans and the environment. Many innovative chemistries have been developed over the past several years that are effective, efficient and more environmentally benign. These approaches include new syntheses and processes as well as new tools for instructing aspiring chemists how to do chemistry in a more environmentally benign manner. The benefits to industry as well as the environment are all a part of the positive impact that Green Chemistry is having in the chemistry community and in society in general.

Over the past few years, the chemistry community has been mobilized to develop new chemistries that are less hazardous to human health and the environment. This new approach has received extensive attention *(1-16)* and goes by many names including Green Chemistry, Environmentally Benign Chemistry, Clean Chemistry, Atom Economy and Benign By Design Chemistry. Under all of these different designations there is a movement toward pursuing chemistry with the knowledge that the consequences of chemistry do not stop with the properties of the target molecule or the efficacy of a particular reagent. The impacts of the chemistry that we design as chemists are felt by the people that come in contact with the substances we make and use and by the environment in which they are contained.

For those of us who have been given the capacity to understand chemistry and practice it as our livelihood, it is and should be expected that we will use this capacity wisely. With knowledge comes the burden of responsibility. Chemists do not have the luxury of ignorance and cannot turn a blind eye to the effects of the science in which we are engaged. Because we are able to develop new chemistries that are more benign, we are obligated to do so.

0097–6156/96/0626–0001$12.00/0

This volume details how chemists from all over the world are using their creativity and innovation to develop new synthetic methods, reaction conditions, analytical tools, catalysts and processes under the new paradigm of Green Chemistry. It is a challenge for the chemistry community to look at the excellent work that has been and continues to be done and to ask the question, "Is the chemistry *I* am doing the most benign that I can make it?".

One obvious but important point: nothing is benign. All substances and all activity have some impact just by their being. What is being discussed when the term benign by design or environmentally benign chemistry is used is simply an ideal. Striving to make chemistry more benign wherever possible is merely a goal. Much like the goal of "zero defects" that was espoused by the manufacturing sector, benign chemistry is merely a statement of aiming for perfection.

Chemists working toward this goal have made dramatic advances in technologies that not only address issues of environmental and health impacts but do so in a manner that satisfies the efficacy, efficiency and economic criteria that are crucial to having these technologies incorporated into widespread use. It is exactly because many of these new approaches are economically beneficial that they become market catalyzed. While most approaches to environmental protection historically have been economically costly, the Green Chemistry approach is a way of alleviating industry and society of those costs.

What is Green Chemistry?

While it has already been mentioned that nothing is truly environmentally benign, there are substances that are known to be more toxic to humans and more harmful to the environment than others. By using the extensive data available on human health effects and ecological impacts for a wide variety of individual chemicals and chemical classes, chemists can make informed choices as to which chemicals would be more favorable to use in a particular synthesis or process. Simply stated, Green Chemistry is the use of chemistry techniques and methodologies that reduce or eliminate the use or generation of feedstocks, products, by-products, solvents, reagents, etc., that are hazardous to human health or the environment.

Green Chemistry is a fundamental and important tool in accomplishing pollution prevention. Pollution prevention is an approach to addressing environmental issues that involves preventing waste from being formed so that it does not have to be dealt with later by treatment or disposal. The Pollution Prevention Act of 1990 *(17)* established this approach as the national policy of United States and the nation's "central ethic" *(18)* in dealing with environmental problems.

There is no doubt that over the past 20 years, the chemistry community, and in particular, the chemical industry, has made extensive efforts to reduce the risk associated with the manufacture and use of various chemicals. There have been innovative chemistries developed to treat chemical wastes and remediate hazardous waste sites. New monitoring and analytical tools have been developed for detecting contamination in air, water and soils. New handling procedures and containment technologies have been developed to minimize exposure. While these areas are laudable efforts in the reduction of risk, they are not pollution prevention or Green

Chemistry, but rather are approaches to pollutant control. Many different ways to accomplish pollution prevention have been demonstrated and include engineering solutions, inventory control and "housekeeping" changes. Approaches such as these are necessary and have been successful in preventing pollution, but they also are not Green Chemistry. There is excellent chemistry that is not pollution prevention and there are pollution prevention technologies that are not chemistry. Green Chemistry is using chemistry for pollution prevention.

No one who understands chemistry, risk assessment and pollution prevention would claim that assessing which substances or processes are more environmentally benign is an easy task. To the contrary, the implications of changing from one substance to another are often felt throughout the life-cycle of the product or process. This difficulty for obtaining a quantifiable measurement of environmental impact has been, however, too often used historically as a rationale for doing nothing. The fact is that for many products and for many processes, clear determinations can be made. Many synthetic transformations have clear advantages over others, and certain target molecules are able to achieve the same level of efficacy of function while being significantly less toxic.

It is important that chemists develop new Green Chemistry options even on an incremental basis. While all elements of the lifecycle of a new chemical or process may not be environmentally benign, it is nonetheless important to improve those stages where improvements can be made. The next phase of an investigation can then focus on the elements of the lifecycle that are still in need of improvement. Even though a new Green Chemistry methodology does not solve at once every problem associated with the lifecycle of a particular chemical or process, the advances that it does make are nonetheless very important.

This volume highlights some of the many advances currently being made in Green Chemistry that are everything from incremental to universal in their impact on the problems that they are addressing. The work described is pioneering and highly innovative, and will provide an information data set of proven Green Chemistry methods and techniques that chemists in the future will need in order to be able to design entire synthetic pathways and processes that are more environmentally benign.

Why is Green Chemistry Important?

In 1993, 30 billion pounds of chemicals were released to air, land and water as tracked by the Toxic Release Inventory of the U.S. Environmental Protection Agency (see Figure 1). While this data covers releases from a variety of industrial sectors, it includes only 365 of the approximately 70,000 chemicals available in commerce today. Of the industrial sectors that are covered by the toxic release inventory, the chemical manufacturing sector is understandably the largest releaser of chemicals to the environment, releasing more than 4 times as many pounds to the environment as the next highest sector (see Figure 2).

The current status of environmental protection in the United States is constructed from a generation of statutes and regulations. The vast majority of these regulations were written at a time where command and control approaches to environmental protection was the order of the day. Many of these laws require

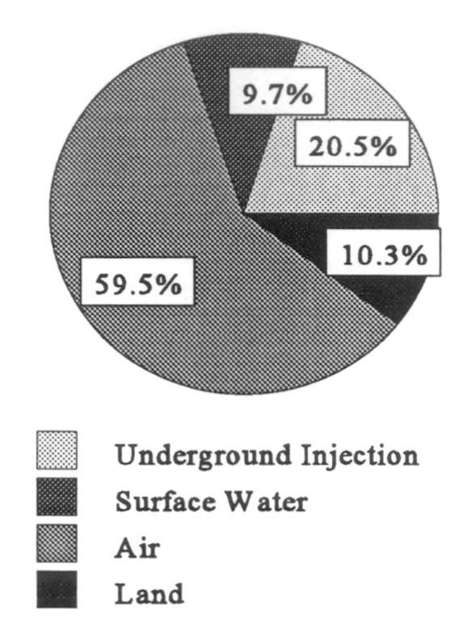

Figure 1. Distribution of Chemical Releases to the Environment

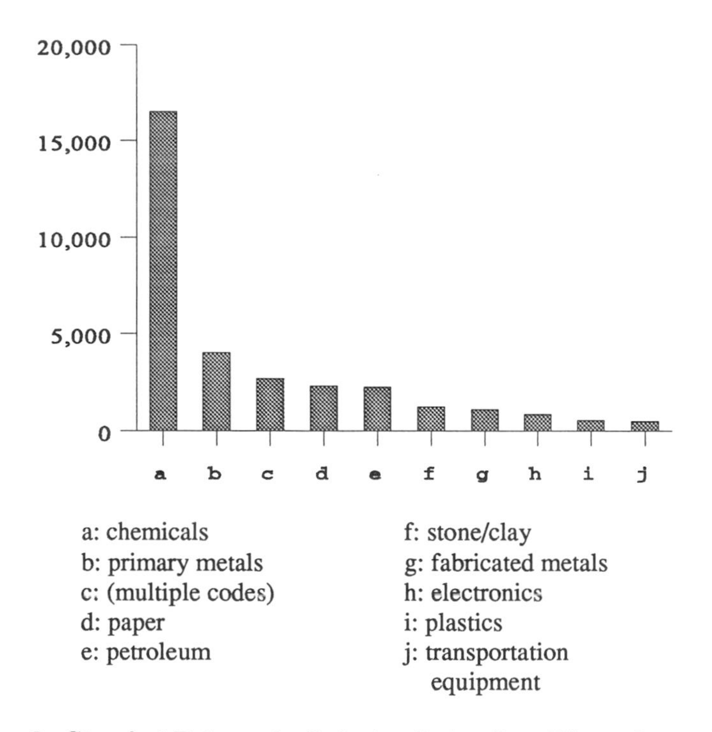

Figure 2. Chemical Releases by Industry Sector (in millions of pounds)

companies either explicitly through methodology-based regulations or implicitly through performance-based regulations to have a variety of waste handling, treatment, control and disposal processes in place to meet environmental mandates. Often these process include equipment with fairly high capital costs.

On a societal level, it is clear that the true costs of the environmental impacts due to the manufacture, processing, use and disposal of all products have not been fully incorporated into the price of the goods. These costs are contained in site remediation, health care expenditures and ecosystem destruction. Therefore, from an economic standpoint, it is clear that we not only want to have sustainable technology but we want it to be cost neutral at a minimum and profitable when at all possible.

The challenge facing industry and society at large is extending technological innovation in a way that is sustainable both economically and environmentally. Certainly the chemical industry has met this challenge economically. In the United States, the chemical industry accounts for the second largest trade surplus of all industrial sectors. With respect to environmental protection, many industrial sectors have made significant progress in reducing emissions over the past decade (see Figure 3). Yet, even with these improvements, the impact of the manufacture, processing, use and disposal of chemicals is staggering.

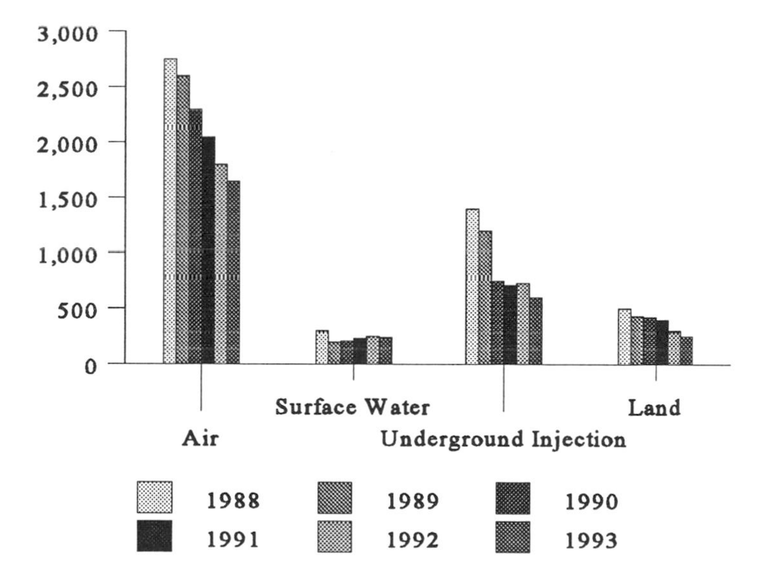

Figure 3. Change in Total Chemical Releases

Green Chemistry provides the best opportunity for manufacturers, processors and users of chemicals to carry out their work in the most economically and environmentally beneficial way. With the challenges facing industry including increased competition, both domestic and international, and increased regulatory requirements, it is clear that every company is searching for new ways of turning cost centers into profit centers now more than ever. It is equally clear that with both transparent and hidden costs within a company, the burden of environmental and hazardous waste operations is one that companies are anxious find creative ways of ridding themselves. This is the promise and potential of Green Chemistry.

In many countries there is talk of the need for an environmental tax where industry would be taxed additionally in order to pay for the environmental impacts that the company has ostensibly caused. Many in industry respond that the cost of complying with environment regulations in the United States is an environmental tax. If this is true, then Green Chemistry can be understood as a tax break waiting to be taken advantage of by industry. By utilizing the principles of Green Chemistry in the design and manufacture of chemical products and processes, companies have reported successes that have dramatically lowered the overall costs associated with environmental health and safety. Environmental expenditures have become to be thought of as the cost of doing business. Green Chemistry is demonstrating new techniques and methodologies which allow industry to continue their tradition of innovation while shifting financial resources that are now expended on environmental costs to further research and development.

In pursuing and developing new Green Chemistry techniques and methodologies as part of the overall efforts to reduce releases of chemical substances to the environment, the scientific community has been among the most creative, as is demonstrated by the work described in this volume.

Areas of Research, Development and Commercialization in Green Chemistry

To more easily understand the advances being made in Green Chemistry, the work being done can be categorized according to what specifically is new or different about a green chemical or process in comparison to the conventional method. A logical breakdown of the chemical reaction results in four basic components:

1. Nature of the Feedstocks or Starting Materials
2. Nature of the Reagents or Transformations
3. Nature of the Reaction Conditions
4. Nature of the Final Product or Target Molecule

It is clear that these four elements are closely inter-related and in some cases inextricably. However, by addressing them separately for the purposes of assessing the potential for designing more environmentally benign syntheses, it is possible to identify the areas where incremental improvements can be made. After identifying an incremental improvement that can be made in an effort to achieve a more environmentally benign synthesis, one must then look at this specific change in the context of the overall synthetic pathway. One must determine whether the new chemistry results in a net improvement to human health and the environment or whether additional incremental improvements are necessary in order to complement the original change in order to ensure that the entire synthetic pathway is more benign.

An assessment of all incremental impacts is necessary to verify the ultimate impact associated with implementation of the new chemistries. This volume highlights some of the recent achievements in Green Chemistry both comprehensive and incremental. Collectively, they are illustrative of the approaches currently being taken in designing environmentally benign methodologies.

Alternative Feedstocks and Starting Materials. The goal of utilizing more environmentally benign feedstocks is to reduce the risk to human health and the environment through a reduction in the hazard of that feedstock. This can be achieved through several different methods including a reduction in the amount of the feedstock used or a reduction of intrinsic toxicity of the feedstock through structural modification or replacement. While risk can be reduced through protective gear and control technologies, in many situations the cost associated with these options makes Green Chemistry solutions very preferable.

One example of feedstock replacement that has been applied commercially is the work of Stern and co-workers *(19-21)* at the Monsanto Corporation in the synthesis of a variety of aromatic amines. By using nucleophilic aromatic substitution for hydrogen, Stern was able to obviate the need for the use of a chlorinated aromatic in the synthetic pathway. Certain chlorinated aromatics are known to be persistent bioaccumulators and to possess other environmental concerns; this research in feedstock replacement resulted in the removal of that concern.

An initiative that is addressing the issue of alternative feedstocks for polymer manufacture is being led by Gross at the University of Massachusetts and centers on the utilization of biological/agricultural wastes such as polysaccharides for the purpose of making new polymeric substances *(22)*. The work is of particular interest because it deals with several environmental concerns simultaneously. It utilizes monomeric materials that are fairly innocuous and thus is an example of the use of environmentally benign feedstocks. The chemistry is also based on a biocatalytic transformation that in several ways offers advantages over many of the reagents conventionally used in the manufacture of polymeric substances. Concern for persistence of polymers and plastics is being addressed by this work as well; Gross' polymers are being designed to biodegrade in post-consumer use.

Significant advances in using alternative feedstocks coupled with biosynthetic methodologies are have been reported from the laboratories of Frost *(23-26)* at Michigan State University. New routes of adipic acid, catechol and hydroquinone have been reported using glucose as a starting material in place of the traditionally used benzene (see Scheme 1). Since benzene is a known carcinogen, a process that would remove it from the synthesis of large volume chemicals in a technically and economically feasible manner is certainly a goal compatible with that of Green Chemistry. The technical advances reported by Frost are examples of such a process.

Replacement of extremely toxic substances such as phosgene is being investigated by several groups. Riley, McGhee, et.al., *(27-35)* at Monsanto Corporation have reported successes in eliminating the use of phosgene in the generation of isocyanates and urethanes through the direct reaction of carbon dioxide with amines. Since phosgene is widely recognized as one of the most acutely toxic substances used in commerce, it would be extremely beneficial with respect to risk reduction to eliminate the need for this substance in the generation of isocyanates. Manzer *(36-37)* at DuPont has reported the development of a catalytic process that again eliminates the use of phosgene from the isocyanate process. In this work the amine is directly carbonylated through the use of carbon monoxide in a proprietary system. The generation of isocyanates by this DuPont process has reportedly been commercialized.

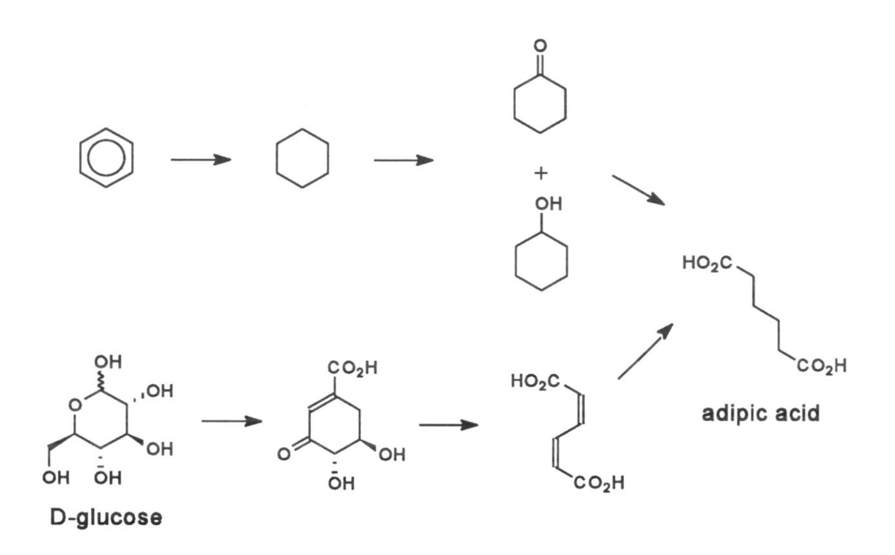

Scheme 1. Conventional and Alternative Syntheses of Adipic Acid

The use of phosgene in the synthesis of polycarbonate polymers is a well known process. Komiya, et.al., *(38)* from Asahi Chemical in Japan have reported the successful generation of polycarbonates that eliminates the use of phosgene in the process by utilizing a molten state reaction between a dihydroxy compound, such as bisphenol A, and a diaryl compound, such as diphenylcarbonate. This approach accomplishes several of the goals of Green Chemistry simultaneously. By eliminating a hazardous substance (phosgene) from the synthetic pathway, risk is reduced. Because the process is conducted in the molten state, the need for the use of methylene chloride, a suspect carcinogen, is also eliminated, thereby further reducing risk.

As is seen in the above examples, many of the approaches that center on the replacement of a hazardous feedstock also address environmental concerns associated with the other elements of the synthetic pathway such as solvents, catalysts, etc. A concern about Green Chemistry that has been expressed is that when one aspect of a synthetic pathway is improved, additional hazards in other parts of the pathway are generated. While this concern needs to be kept in mind when evaluating a Green Chemistry process, the exact opposite is most often observed. When chemistry is designed to minimize a concern in one part of a synthetic pathway, it often fortuitously reduces the hazards in other parts of the pathway as is demonstrated by the above examples.

Alternative Synthetic Transformations and Alternative Reagents. In the goal to reduce risk to human health and the environment through the elimination or reduction of toxic substances, there are significant opportunities to substitute more benign chemicals for the reagents that are necessary to carry out particular transformations or change the actual transformations themselves.

The use of an alternative methodology to accomplish transformations important in the pharmaceutical and dye industries has been reported by Epling and co-workers *(39)* at the University of Connecticut. Through the use of visible light as the "reagent", Epling accomplishes the cleavage of a variety of dithiane and oxathiane ring systems commonly used as protecting groups (see Table I). Traditionally, rings of this type are most commonly cleaved by heavy metal catalyzed reduction. Through this new methodology, Epling has introduced an additional synthetic option for consideration that does not possess the environmental and health concerns associated with the use of heavy metals.

Another use of visible light that eliminates the use of substances with environmental concerns has been reported by Kraus *(40-41)* from Iowa State University. The transformations investigated by Kraus involves the widely used Friedel-Crafts Acylation reaction which is catalyzed by Lewis Acids. By using quinolic moieties and aldehydes, Kraus was able to achieve formal synthesis of a number of target molecules of importance in the pharmaceutical industry including diazapam and analogues.

Methylation is a transformation that is extremely important in the manufacture of a variety of chemicals but also is potentially very hazardous to human health. The health hazards are derived primarily from the fact that many strong methylating agents, such as dimethyl sulfate, possess very high acute toxicity and are often carcinogenic. Tundo *(42-43)*, at the University of Venice, discusses in a later chapter of this volume the use of dimethylcarbonate as a replacement for dimethylsulfate in a variety of methylations. The results reported here focus mainly on the use of dimethylcarbonate in the methylation of arylacetonitriles and methyl arlyacetates which are important in the synthesis of a class of anti-depressant drugs. Again, this advance in Green Chemistry results in not only the elimination of the toxic reagent dimethyl sulfate but also the virtual elimination of the problem of high salt production during the transformation because the process requires only a catalytic amount of base to proceed.

In investigating the replacement of dimethyl sulfate with dimethyl carbonate in some applications, a reasonable question is how the dimethyl carbonate is prepared since traditionally, dimethylcarbonate is prepared by the reaction of phosgene with methanol. The overall benefit to human health and the environment is questionable if the use of dimethyl carbonate merely replaces one extremely toxic substance, dimethylsulfate, with another, phosgene. This, however, is not the case due to the work discussed by Rivetti and co-workers *(48-53)* in a later chapter in this volume. In this chapter, Rivetti describes the approach that the Enichem chemical company is

Table I. Isolated Yields of Aldehydes or Ketones from Dye-Promoted Photocleavage of Dithio Compounds

Entry	Substrate	Product	Isolated Yield
1	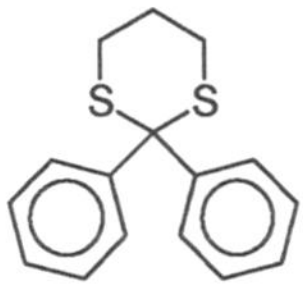	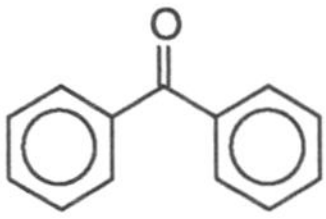	97%
2	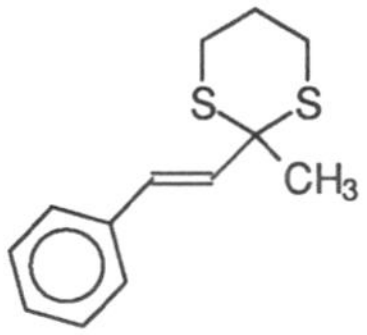	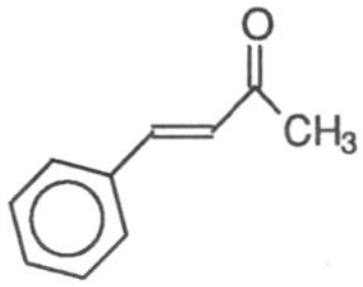	94%
3	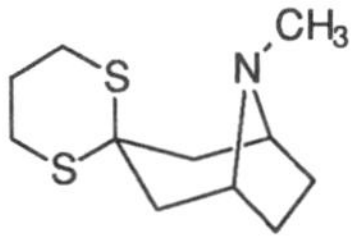	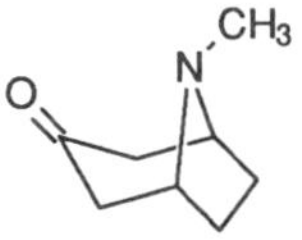	95%
4	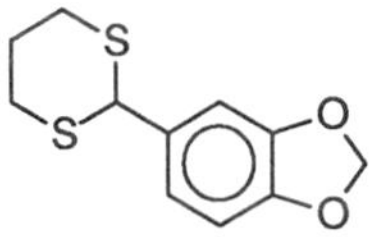	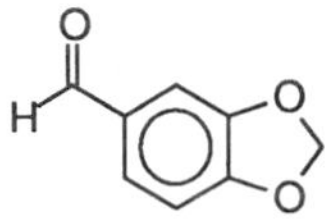	91%
5	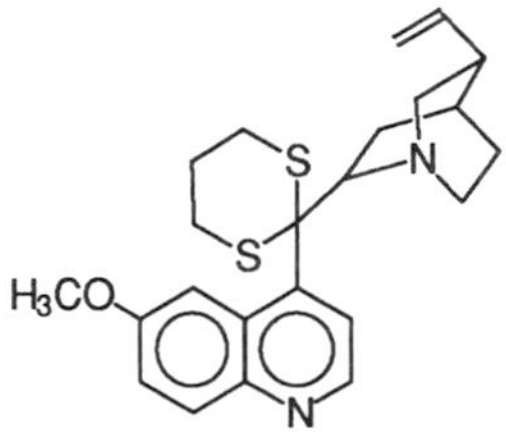	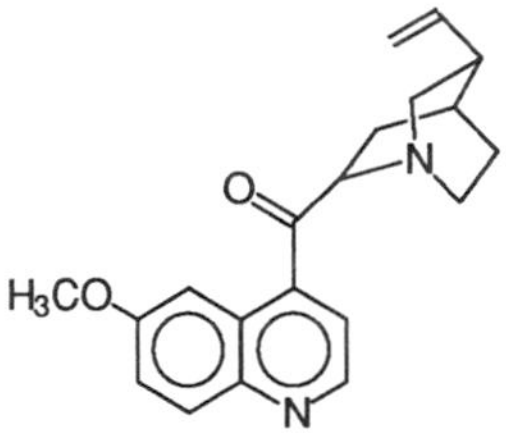	86%
6	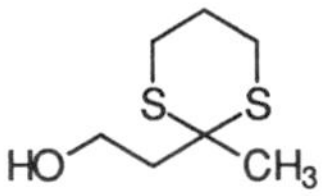	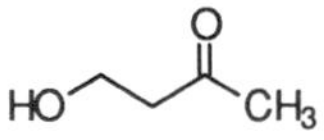	91%
7	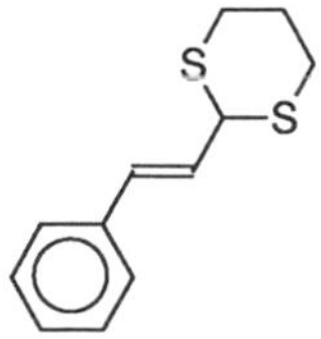	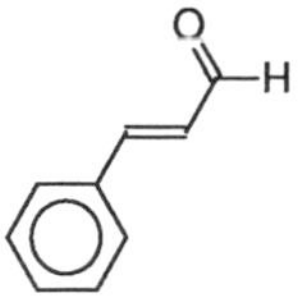	95%

taking in the production of dimethyl carbonate, namely the oxidative carbonylation of methanol using carbon monoxide.

In an approach that differs significantly from those mentioned above, Paquette *(54)* of Ohio State University utilizes indium in effecting transformations that demonstrate both increased selectivity and a decreased use of volatile organic solvents. Paquette's investigations currently use this relatively non-toxic metal to catalyze various reactions in place of metals of greater environmental concern. In addition, the transformations are able to be performed in an aqueous solvent which eliminates the need for various VOC solvents that are of concern for both air and water pollution.

Alternative Reaction Conditions. The conditions that are used in synthesizing a chemical can have a significant effect on the pathway's overall environmental impact. Considerations of the amount of energy used by one process versus another is quite easily evaluated in economic terms but is not currently as easily evaluated in environmental terms. It appears that because of this difficulty in measurement, and not necessarily as a judgement on relative importance, that the majority of Green Chemistry research on reaction conditions has been centered around the substances utilized as part of those conditions.

A great deal of the environmental concern associated with chemical manufacture comes from not merely the chemicals that are made or the chemicals from which they are made, but from all of the substances associated with their manufacture. In the manufacture, processing, formulation and use of chemical products, there are a variety of associated substances that contribute to the environmental loading from chemical manufacture. The most visible of these associated substances are the solvents used in reaction media, separations and formulations. Many solvents, especially the widely used volatile organic solvents, have come under increased scrutiny and regulatory restriction based on concerns for their toxicity and their contributions to air and water pollution. It is for these economic and environmental reasons that much of the research and development in Green Chemistry reaction conditions is focussed on alternative solvents.

One of the most active areas of investigation in alternative solvents and in Green Chemistry in general has focused on the use of supercritical fluids (SCFs) as solvents. Solvent systems such as supercritical (SC) carbon dioxide and SC carbon dioxide/water mixtures are being investigated systematically for their usefulness in a wide range of reaction types. In general, SCFs appear to offer the promise of providing a low cost, innocuous solvent that can supply "tuneable" properties depending on where in the critical region one decides to conduct their chemistry. Current research is focussing, in part, on determining the range of applications in which SCFs can be used and on outlining the limitations associated with SCF use in order to assess the future role of supercritical fluids in synthetic chemistry. There have been studies over the years on supercritical fluids that provide the basis for a physical chemistry profile of this class. While there are anecdotal reports that various industrial interests ave examined the use of SCFs for specific applications, most of this work was not published in the open scientific literature. Therefore, references to the use of SCFs as a reaction medium for chemical synthesis are scarce.

The early studies carried out during this resurgence of interest in supercritical fluids have been quite promising. The studies conducted by Tanko *(55-58)* at Virginia Polytechnic Institute and State University on free-radical reactions in SC-CO_2 provided a foundation for understanding how a classical reaction type would function in this new solvent system. Using standard bromination reactions of alkylated aromatics as the model system, Tanko demonstrated that the yields and selectivities of free-radical halogenations in supercritical fluids were equal and in some cases superior to those conducted in conventional solvent systems. As is the case in all Green Chemistry alternatives, demonstrating technical efficacy is crucial in order to evaluate the true advantages offered by the environmental and risk reduction benefits.

Polymerization reactions in supercritical fluids have been studied extensively in the laboratory of DeSimone *(59-61)* at the University of North Carolina. DeSimone had demonstrated the ability to synthesize a variety of polymer types with several different monomeric systems. His methyl methacrylate polymer studies have demonstrated that there are pronounced advantages to using supercritical fluids as a solvent system compared to using conventional halogenated organic solvents.

The National Laboratory at Los Alamos has been actively engaged over the past several years in research in the applications of SC-CO_2 as a synthetic solvent. The work of Tumas *(61-66)* and co-workers as detailed in a later chapter of this volume profiles the performance of reactions such as polymerization of epoxides, oxidation of olefins and asymmetric hydrogenations in supercritical systems. In each of these cases the reactions proceeded without compromise when compared to conventional solvent systems, and superior performance was reported in the asymmetric hydrogenation reactions.

Although currently under extensive study, supercritical fluids are the only example of an alternative solvent under investigation for Green Chemistry. Hatton *(67)* at MIT has reported early results of the use of amphiphilic star polymers as solvents in synthesis. In addition to their innocuous nature, due in part to their size, they also have the advantage of minimizing the need for intensive separations that can require additional solvent use.

Too numerous to list are the projects that are currently investigating the use of aqueous solvent systems in place of organic solvents in chemical manufacturing. While many of these show great promise and in some cases improved efficacy, the environmental impact is one that needs to be carefully evaluated on a case-by-case basis. It is possible that what was once an air pollution problem brought on by the use of volatile organic compounds could become a more serious water pollution problem if wastes are difficult to remove from the aqueous solvent system and are subsequently lost as effluent. Even though the environmental concern associated with a conventional process is eliminated by the use of an aqueous solvent system, the alternative process cannot be considered a viable substitute until the total impacts of the substitution is assessed.

Alternative Products and Target Molecules. The goal of designing safer chemicals is both straightforward and extremely complex. It is well recognized that in many cases the part of a molecule which provides its intended activity or function is separate from the part of a molecule responsible for its hazardous properties or toxicity.

Therefore, the challenge is to reduce the toxicity of a molecule without sacrificing the efficacy of function. This goal has been actively pursued in specific industrial sectors, such as pharmaceutical and pesticide manufacture, where it is of obvious importance. With some notable exceptions, this same goal has been generally ignored in most of the other chemical sectors.

The following examples provide insight into how an understanding of the mechanism of action of the toxicity of chemical coupled with the knowledge of the structural requirements for the desired function can be used to design safer chemicals. By utilizing the knowledge that synthetic chemists possess, it is possible to greatly reducing the hazards associated with certain chemical products while maintaining their performance and innovation that has become expected from the chemical industry.

The work conducted by DeVito *(68)* at the U.S. Environmental Protection Agency on a large range of nitriles elucidated the nature of the hazard posed by the release of cyanide, since the degree of toxicity can be directly correlated to the ability of the nitrile to form an alpha radical. By blocking the alpha position such that radical formation is not possible, the nitrile's toxicity can be decreased by several orders of magnitude without adversely effecting the ability of the nitrile to carry out its function as a cross-linking agent.

Another example of designing chemicals such that their function is maintained without the toxicity is the work of DePompei *(69)* at Tremco, Inc., in designing alternatives to isocyanates as sealants. This work utilizes acetoacetate esters which as a class do not have the human health and environment concerns that are associated with isocyanates in this use. Rather than modifying the molecular structure in order to reduce the hazard, this work demonstrates the approach of reexamining the function that needs to be performed and subsequently identifying or developing compounds that can accomplish this function without the accompanying hazards.

Green Chemistry's approach to designing safer chemicals is simply another option for chemists to employ in their overall evaluation and decision making process on what to make and how to make it. A great deal of effort has been expended in developing ways of handling hazardous chemicals in a safe manner such that the risk of injury and adverse health effects is minimized. In many cases this approach to minimizing risks is a costly one. Where appropriate, chemicals can be designed such that their hazards are already minimized thereby obviating the need for expenditures on handling and control procedures.

Universal Issues. While it is possible to classify the Green Chemistry research that is currently being conducted into several categories, there are some subjects that do not fit neatly into these categories because of their overarching nature. The work being done in these areas is both innovative and significant and more importantly, fundamental to the basic direction that Green Chemistry needs to follow in order to be successful in all its endeavors.

At the foundation of the principles of Green Chemistry is that of Atom Economy. This concept as elucidated and demonstrated by Trost *(70)* couples the most important elements of the idea of environmentally benign synthesis and the definition of synthetic elegance to provide a framework for how synthetic design should be guided. The economical use of atoms in the construction of a synthetic pathway, avoiding the use of blocking, protecting or leaving groups whenever possible, is at the core of providing the most waste-free process and environmentally benign synthesis. It is true that the principle of Atom Economy does not directly address the issue of hazard or toxicity. However, the incorporation of the goals of this paradigm is and needs to be one of the central tenants for any synthesis that is striving toward Green Chemistry.

Another fundamental area of investigation that can effectively accomplish the goals of Green Chemistry is that of catalysis. While this volume addresses the topic in a later chapter *(71-72)*, it can only hope to scratch the surface of such a broad area of investigation. Several volumes could be written on the research that is being done in both industry and academia on catalysis as it relates to environmentally benign chemistry. Asymmetric catalysis, solid acid catalysis, biocatalysis, heterogeneous catalysis and phase transfer catalysis are just a few examples that all have a direct and significant impact on accomplishing the goals of Green Chemistry. A truly fundamental question of whether a reaction must proceed stoichiometrically or can be conducted catalytically is one of great import for the ultimate nature of the reaction as whole and for the determination of the extent to which it is environmentally benign.

Conclusions

In the history of chemistry, there have been a number of periods where the chemistry community as a whole or sections of the community focussed on a goal. These goals have included everything from synthesizing classes of chemicals (e.g. natural products) to developing reaction types (stereo-specific reactions) to defining applications (e.g. anti-cancer agents). Often the goal, as stated, merely provides an objective on which to focus energy and resources. The pursuit of the goal often provides the avenue for many other accomplishments as well.

When goals such as those mentioned above are considered to be important to the field of chemistry and to society in general, a higher level of funding, support and recognition becomes available for research in those areas. Dramatic increases in funding for research in Green Chemistry has been seen over the past several years. With the high quality of the research demonstrated thus far in Green Chemistry and the economic benefits that have been and continue to be realized, the support is expected to continue to increase.

Not every research proposal whose goal was to synthesize natural products actually produced one, and not every research project which strove to make a chemotherapeutic agent succeeded in achieving its objective. However, excellent chemistry has nonetheless resulted from such research. The same is true of Green Chemistry. Not every project is going to achieve its goal of innocuous feedstocks or reagents, or benign conditions or products, but in striving for this necessary and worthwhile goal, even these projects will, and in fact have, resulted in excellent chemistry.

Literature Cited

1. Hall, N. *Science* **1994**, *266*, pp. 32-34.
2. Newman, A. *Environ. Sci. Technol.* **1994**, *28(11)*, p. 463A.
3. Wedin, R. *Today's Chemist at Work* **1994** (June), pp. 114-18.
4. Sheldon, R.A. *Chemtech* **1994** (March), pp. 38-47.
5. Amato, I. *Science* **1993**, *259*, pp. 1538-1541.
6. Illman, D.L. *Chem. Eng. News* **1993** (Sept. 6), pp. 26-30.
7. Illman, D.L. *Chem. Eng. News* **1993** (March 29), pp. 5-6.
8. Wedin, R. *Today's Chemist at Work* **1993** (July/Aug.), pp. 16-19.
9. Ember, L. *Chemtech* **1993** (June), p. 3.
10. Miller, J.P. *The Wall Street Journal* **1993** (Dec. 7), p. B6.
11. Crenson, M. *Dallas Morning News* **1993** (Sept. 13), p. 1D.
12. Woods, M. *Sacramento Bee* **1993** (August 26), p. A18.
13. *Business Week* **1993** (August 30), *3334*, p. 65.
14. Rotman, D. *Chem. Week* **1993** (Sept. 22), *153(10)*, pp. 56-57.
15. Mitchell, J.P. *Proc. Natl. Acad. Sci. USA* **1992**, *89*, pp. 821-826.
16. Ember, L. *Chem. Eng. News* **1991** (July 8), pp. 7-16.
17. Pollution Prevention Act of 1990. 42 U.S.C.§§ 13101-13109, **1990**.
18. Browner, C.M. *EPA Journal* **1993**, *19(3)*, pp. 6-8.
19. Stern, M.K. In *Benign By Design: Alternative Synthetic Pathways for Pollution Prevention;* Anastas, P.T.; Farris, C.A., Eds.; ACS Symposium Series #577; American Chemical Society: Washington, D.C., **1994**, pp. 133.
20. Stern, M.K.; Hileman, F.D.; Bashkin, J.K. *J. Am. Chem. Soc.* **1992**, *114*, pp. 9237-9238.
21. Stern, M.K.; Cheng, B.K. *J. Org. Chem.* **1993**, *58,* pp. 6883-6888.
22. Gross, R.A.; Kim, J.H.; Gorkovenko, A.; Kaplan, D.L.; Allen, A.L.; Ball, D. In *Preprints of Papers Presented at the 208th ACS National Meeting;* Bellen, G.E., Chairman; Division of Environmental Chemistry, American Chemical Society: Washington, D.C., **1994**, *34(2)*, pp. 228-229.
23. Draths, K.M.; Frost, J.W. In *Benign By Design: Alternative Synthetic Pathways for Pollution Prevention;* Anastas, P.T.; Farris, C.A., Eds.; ACS Symposium Series #577; American Chemical Society: Washington, D.C., **1994**, pp. 32.
24. Draths, K.M.; Frost, J.W. *J. Am. Chem. Soc.* **1994**, *116*, p. 399.
25. Draths, K.M.; Frost, J.W. *J. Am. Chem. Soc.* **1990**, *112*, p. 9630.
26. Draths, K.M.; Ward, T.L.; Frost, J.W. *J. Am. Chem. Soc.* **1992**, *114*, p. 9725.
27. Riley, D. In *Benign By Design: Alternative Synthetic Pathways for Pollution Prevention*; Anastas, P.T.; Farris, C.A., Eds.; ACS Symposium Series #577; American Chemical Society: Washington, D.C., **1994**, pp. 122.
28. See Chapter in this book by McGhee, W.D.; Doster, M.; Riley, D.; Ruettimann, K.; Solodar, J.; and Waldman, T.
29. McGhee, W.D.; Riley, D.P. *Organometallics* **1992**, *ll*, pp. 900-907.

30. McGhee, W.D.; Riley, D.P.; Christ, M.E.; Christ, K.M. *Organometallics* **1993**, *12*, pp. 1429-1433, and references cited therein.
31. Riley, D.P.; McGhee, W.D. U.S. Patent # 5,055,577, **1991**.
32. Riley, D.P.; McGhee, W.D. U.S. Patent # 5,200,547, **1993**.
33. McGhee, W.D.; Parnas, B.L.; Riley, D.P.; Talley, J.J U.S. Patent # 5,223,638, **1993**.
34. McGhee, W.D.; Stern, M.K.; Waldman, T.E. U.S. Patent # 5,233,010, **1993**.
35. McGhee, W.D.; Waldman, T.E. U.S. Patent # 5,189,205, **1993**.
36. Manzer, L.E. In *Benign By Design: Alternative Synthetic Pathways for Pollution Prevention;* Anastas, P.T.; Farris, C.A., Eds.; ACS Symposium Series #577; American Chemical Society: Washington, D.C., **1994**, pp. 144.
37. (DuPont) U.S. Patent # 4,537,726, **1985**.
38. See Chapter in this book by Komiya, K., et.al.
39. Epling, G.A.; Wang, Q. In *Benign By Design: Alternative Synthetic Pathways for Pollution Prevention;* Anastas, P.T.; Farris, C.A., Eds.; ACS Symposium Series #577; American Chemical Society: Washington, D.C., **1994**, pp. 64.
40. Kraus, G.A.; Kirihara, M.; Wu, Y. In *Benign By Design: Alternative Synthetic Pathways for Pollution Prevention;* Anastas, P.T.; Farris, C.A., Eds.; ACS Symposium Series #577; American Chemical Society: Washington, D.C., **1994**, pp. 76.
41. Kraus, G.A.; Kirihara, M. *J. Org. Chem.* **1992**, *57*, p. 3256.
42. See Chapter in this book by Tundo, P.; Selva, M.; Marques, C.A.
43. Tundo, P.; Selva, M. *Chemtech* **1995**, pp. 31-35.
44. Tundo, P.; Trotta, F.; Moraglio, G. *J. Chem. Soc., Perkin Trans.* I **1989**, p. 1070.
45. Tundo, P.; Trotta, F.; Moraglio, G. *J. Org. Chem* **1987**, *52*, p. 1300.
46. Tundo, P.; Trotta, F.; Moraglio, G.; Ligorati, F. *Ind. Eng. Chem. Res.* **1988**, *27*, p. 1565.
47. Loosen, P.; Tundo, P.; Selva, M. U.S. Patent # 5,278,533, **1994**.
48. See Chapter in this book by Rivetti, F.; Delledonne, D.
49. Rivetti, F.; Romano, U. U.S. Patent # 5,206,409(ECS).
50. Rivetti, F.; Romano, U. U.S. Patent # 5,159,099(ECS).
51. Rivetti, F.; Romano, U.; DiMuzio, N. U.S. Patent # 4,318,862(ECS).
52. Delledonne, D.; Rivetti, F.; Romano, U. EP Pat. Appl. 463,678(ECS).
53. Delledonne, D.; Rivetti, F.; Romano, U. *J. Organomet. Chem.* **1995**, *488*, p. C15.
54. Paquette, L. In *Abstracts of Papers Presented at the 209th ACS National Meeting*; Pasto, D.J., Program Chairperson; Division of Organic Chemistry, American Chemical Society: Washington, D.C., **1995**, #125.
55. Tanko, J.M.; Blackert, J.F.; Sadeghipour, M. In *Benign By Design: Alternative Synthetic Pathways for Pollution Prevention;* Anastas, P.T.; Farris, C.A., Eds.; ACS Symposium Series #577; American Chemical Society: Washington, D.C., **1994**, p. 98.
56. Tanko, J.M.; Blackert; J.F. *Science* **1994**, *263*, p. 203.
57. Tanko, J.M.; Mas, R.H.; Suleman, N.K. *J. Am. Chem Soc.* **1990**, *112*, p. 5557.

58. Tanko, J.M.; Suleman, N.K.; Hulvey, G.A.; Park, A.; Powers, J.E. *J. Am. Chem. Soc.* **1993**, *115*, p. 4520.
59. DeSimone, J.M.; Guan, Z.; Elsbernd, C.S. *Science* **1992**, *257*, p. 945.
60. DeSimone, Y.M.; Maury, E.E.; Lemert, R.E.; Combes, J.R. *Polym. Mater. Sci. Eng.* **1993**, *68*, p. 41.
61. DeSimone, J.M.; Maury, E.E.; Menceloglu, Y.Z.; McClain, J.B.; Romack, T. J.; Combes, J.R. *Science* **1994**, *265*, p. 356.
62. Burk, M.J.; Feng, S.; Gross, M.F.; Tumas, W. J. *J. Am. Chem. Soc.* **1995**, *117*, p. 8277.
63. Morgenstern, D.A.; Tumas, W. *J. Am. Chem. Soc.* submitted.
64. Burk, M.J.; Harper, G.P.; Kalberg, C.S. *J. Am. Chem. Soc.* **1995**, *117*, p. 4423.
65. Burk, M.J. *J. Am. Chem. Soc.* **1991**, *113*, p. 8518.
66. Burk, M.J.; J.E., F., Nugent, W.A.; Harlow, R.L. *J. Am. Chem. Soc.* **1993**, *115*, p. 10125.
67. Hurter, P.N.; Hatton, T.A. *Langmuir* **1992**, *8*, pp. 1291-1299.
68. Grogan, J., DeVito, S.C.; Pearlman, R.S.; Korzekwa, K.R.; Modeling Cyanide Release From Nitriles: Prediction of Cytochrome P450 Mediated Acute Nitrile Toxicity. *Chem. Res. in Tex.* **1992**, *5*, pp. 548-552.
69. DePompei, M.F. In *Preprits of Papers Presented at the 208th ACS National Meeting*; Bellen, G.E., Chairman; Division of Environmental Chemistry, American Chemical Society: Washington, D.C., **1994**, *34(2)*, p. 393.
70. Trost, B.M. *Science* **1991**, *2*, pp. 1471-1477.
71. See Chapter in this book by Simmons, M.
72. Cusamano, J. *Chemtech* **1992**, pp. 482-489.

RECEIVED December 14, 1995

Alternative Feedstocks and Starting Materials

Chapter 2

New Process for Producing Polycarbonate Without Phosgene and Methylene Chloride

Kyosuke Komiya[1], Shinsuke Fukuoka[1], Muneaki Aminaka, Kazumi Hasegawa, Hiroshi Hachiya, Hiroshige Okamoto, Tomonari Watanabe[2], Haruyuki Yoneda[3], Isaburo Fukawa[3], and Tetsuro Dozono[4]

Chemisty & Chemical Process Laboratory, Asahi Chemical Industry Co., Ltd., Kojima-Shionasu, Kurashiki-City, Okayama 711, Japan

A new polymerization technology for manufacturing polycarbonates without using phosgene and methylene chloride has been established. The innovative and environmentally benign process is based on the entirely new concept, "Solid-State Polymerization of Amorphous Polymers". Asahi's new process essentially consists of (1) prepolymerization, (2) crystallization, and (3) solid-state polymerization. The solid-state polymerization process is able to cover a very wide range of molecular weight polycarbonates from the lowest disk grade (Mw 15,000) to the ultra high molecular weight grade (Mw>60,000), which is very difficult to produce by conventional processes. The polycarbonates obtained by Asahi's new process are colorless with good transparency; other excellent features include, for example, better heat stability and reworkability than that of polycarbonates produced by the phosgene process. Furthermore, the ultra high molecular weight polycarbonate produced by the new process also has excellent properties, which are missing in the conventionally available polycarbonate, such as excellent solvent resistance and excellent steam resistance.

It is well known that phosgene is currently used industrially in large scales in the two most important processes for manufacturing polycarbonates and isocyanates all over the world. Phosgene, however, is notorious for its high toxicity (ACGIH TLV-TWA 0.1 ppm) and corrosiveness. Therefore, environmentally benign processes, which are able to be commercialized and that do not require phosgene have been desired earnestly for a long time.

[1]Corresponding authors
[2]Current address: Division of Chemicals Technology, Yakou, Kawasaki-City, Japan
[3]Current address: Department of Research and Development Administration, Hibiya-Mitsui Building 1–2, Yurakucho, 1-chome, Chiyoda-ku, Tokyo, Japan
[4]Current address: Fuji Office, Samejima, Fuji-City, Japan

0097–6156/96/0626–0020$12.00/0

Asahi Chemical Industry Co., Ltd. has succeeded in developing alternative and innovative non-phosgene processes for producing isocyanates and polycarbonates in the pilot scales which are commercially viable. In the production of isocyanates, processes for both aromatic isocyanates, such as methylene diphenyl diisocyanate (MDI), and aliphatic isocyanates, such as hexamethylene diisocyanate (HDI) or isophorone diisocyanate (IPDI), have already been developed successfully. A part of those processes has already been reported (*1*), and the others will be reported in the near future. In this paper, the new innovative process for producing polycarbonates is reported.

Polycarbonates are well known to be typical amorphous polymers and to have excellent properties such as heat resistance, impact resistance, transparency, and dimensional stability (*2-5*). Polycarbonates, therefore, have been widely employed in various applications from nursing bottles to precision instruments (CDs, cameras, etc.), or in structural materials (for electrical applications, electronics, automobiles, construction applications, etc.). The global demand of polycarbonates has been growing more than 10% per year. The production capacity of polycarbonate world wide is about 1 million tons per year, and the boom in polycarbonate plant construction continues. Almost all of the polycarbonates, however, have been produced by the "Phosgene Process".

"Phosgene Process". In the phosgene process, the polycarbonate is produced by an interfacial condensation polymerization of bisphenol-A with phosgene and NaOH between two solvents, methylene chloride and water (Figure 1). The obtained methylene chloride solution of the crude polycarbonate is washed with water to remove the by-product, NaCl, but the washed aqueous solution contains not only NaCl but also methylene chloride. The methylene chloride in the polycarbonate solution is also removed, but the complete removal is difficult because methylene chloride has a strong affinity to polycarbonate. Polycarbonates produced by the phosgene process, therefore, typically contain chlorine impurities which have a negative effect on polymer properties.

Although the most noted problem associated with this process is the use of phosgene, the process has another significant problem, namely the process must use a very large amount (more than about 10 times by weight of the polycarbonate to be produced) of methylene chloride. Methylene chloride is a toxic chemical (IARC group-2B, possible carcinogenic to humans; EPA group-B2, probable human carcinogen) and is also one of the 17 chemicals targeted for emissions reduction by EPA, known as the 33/50 Program (an EPA voluntary pollution prevention initiative). In the phosgene process, the recovery of methylene chloride can be costly, due to its low boiling point (40°C) and its high solubility in water (20g/l).

"Melt Process". In the "Melt Process", which has been in development for a long time without success (*6*), a polycarbonate is produced by performing a molten-state ester exchange reaction between bisphenol-A and diphenyl carbonate in the presence of a catalyst, while eliminating phenol (Figure 2). However, in order to attain the desired degree of polymerization by this process, phenol and, subsequently, diphenyl carbonate need to be distilled from a formed molten polycarbonate of extremely high viscosity, which is very difficult. In addition, it is generally necessary to perform the reaction at a temperature as high as 280°C to 310°C under a high vacuum of 1 mm Hg or less for a long residence time. Consequently, this process has many disadvantages, such as discoloration of the polymer due to long residence at high temperature and high vacuum and difficulty in producing a polymer with the molecular weight necessary for structural use(*4,7*).

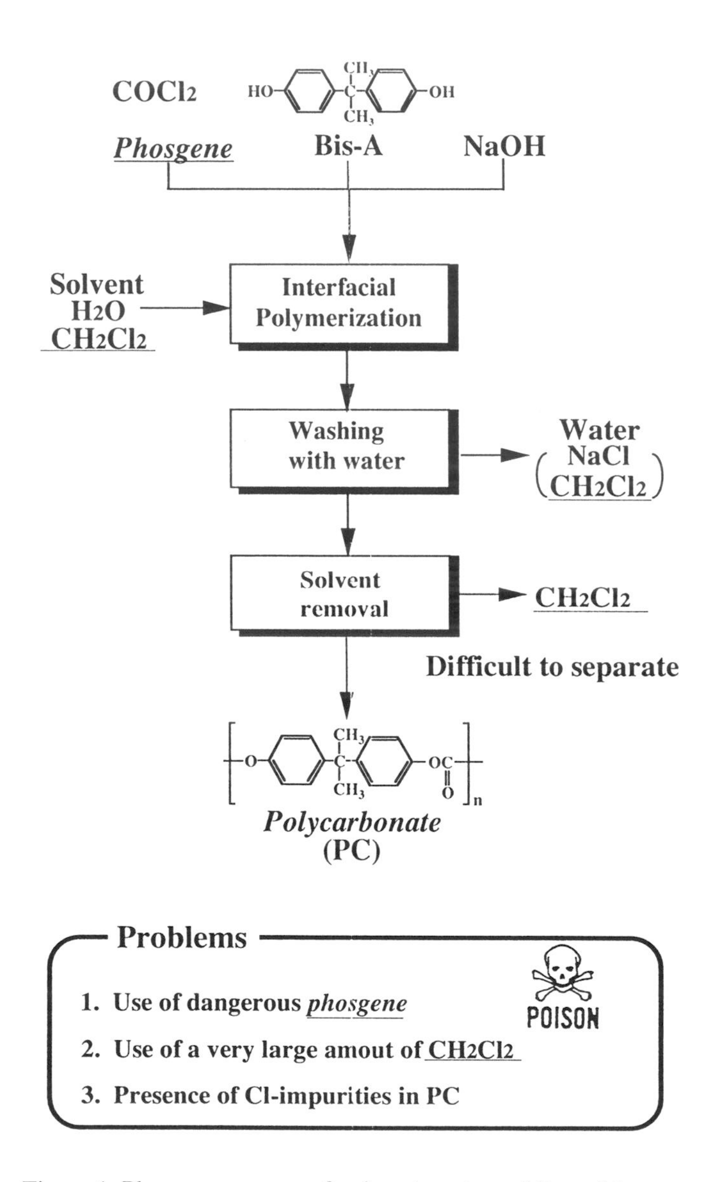

Figure 1. Phosgene process of polycarbonate, and its problems.

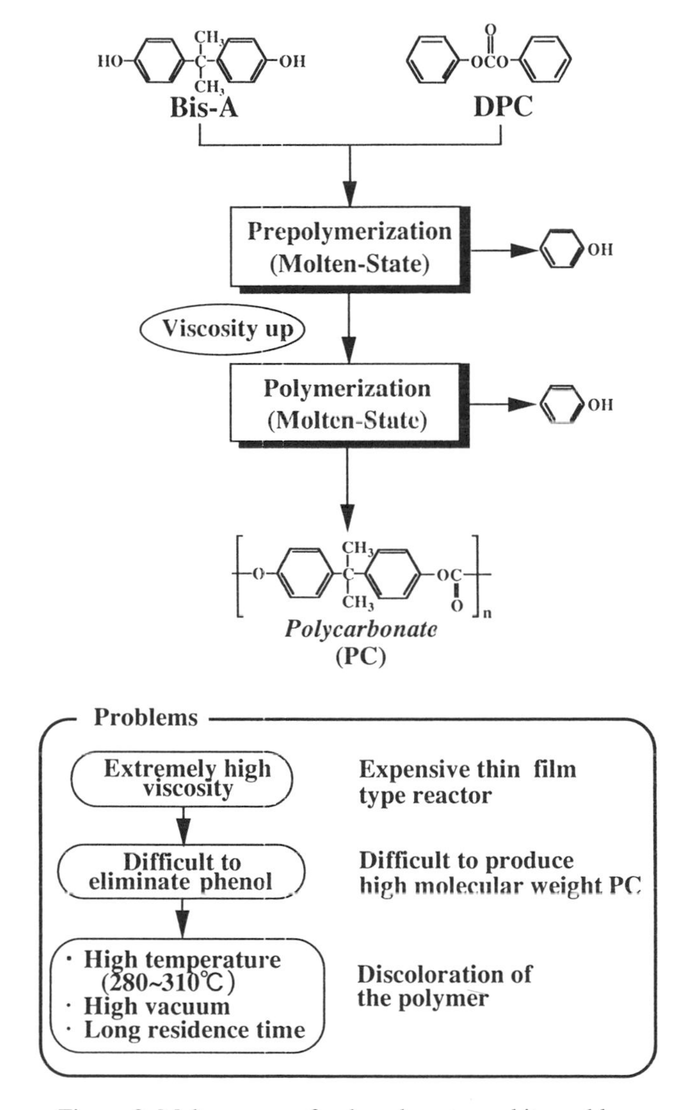

Figure 2. Melt process of polycarbonate, and its problems.

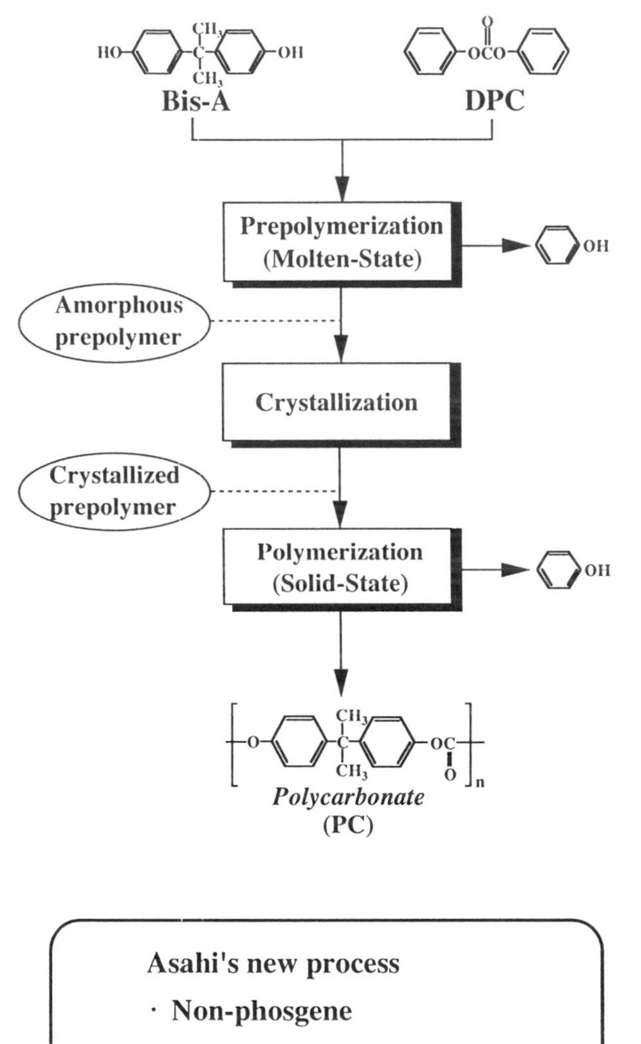

Asahi's new process

- **Non-phosgene**
- **Non-methylene chloride**

Figure 3. Solid-state polymerization process of polycarbonate

Raw materials of Asahi's New Process. In Asahi's new process, an aromatic dihydroxy compound, such as bisphenol-A, and a diaryl carbonate, such as diphenyl carbonate, are also used as raw materials. Although diphenyl carbonate is conventionally produced by the reaction of phenol with phosgene, Asahi Chemical has succeeded in developing an innovative process for producing diphenyl carbonate from dimethyl carbonate without phosgene (*8,9*), a process thought to be difficult to develop. Furthermore, a process for producing dimethyl carbonate from an alkylene carbonate and methanol have been developed (*10*). As the alkylene carbonate is produced from the alkylene oxide and CO_2, the carbonate group source of the polycarbonate obtained by Asahi's process is CO_2. Details of these processes, producing diphenyl carbonate and dimethyl carbonate, will be reported in other papers.

Asahi's Solid-state Polymerization Process

A new polymerization technology for manufacturing polycarbonate has been established without using phosgene and methylene chloride (*11,12*). The new process, called "Solid-state Polymerization Process", consists of three steps, namely, prepolymerization, crystallization, and solid-state polymerization (Figure 3). In the prepolymerization step, the amorphous prepolymer is obtained by molten-state prepolymerization of bisphenol-A and diphenyl carbonate. The amorphous prepolymer is converted to the crystallized prepolymer in the crystallization step, and finally, in the solid-state polymerization step, the polycarbonate of the desired molecular weight is obtained.

Although solid-state polymerizations of polyamides and polyesters (which are crystalline polymers), have been known since 1939 and 1962 (*13,14*), until now, it has been considered impossible to produce polycarbonate by solid-state polymerization, because polycarbonates are amorphous polymers and become molten at the temperatures necessary to effect polymerization. The key technology in solid-state polymerization of polycarbonate is the crystallization of the amorphous prepolymer. It has been found that the low molecular weight amorphous prepolymer is easily crystallized, and the obtained crystallized prepolymer retains its solid-state when it is heated to the temperatures necessary for polymerization.

Process Description by Step

Prepolymerization Step. A clear amorphous prepolymer is obtained by performing a molten-state prepolymerization between bisphenol-A and diphenyl carbonate, while eliminating phenol (Figure 4). In this step, obtaining a low molecular weight (Mw 2,000~20,000) prepolymer with low melt viscosity is sufficient, therefore, the prepolymerization can be easily carried out at a relative low temperature (< 250°C). Discoloration of the prepolymer does not occur because of low temperature and short residence time compared to the melt process which must be performed at a temperature as high as 280°C to 310°C.

Crystallization Step. The crystallization step gives a crystallized prepolymer with high melting point that is sufficient for performing the solid state polymerization (Figure 5). It has been found that the clear, amorphous, low molecular weight prepolymer can be easily converted to a white, opaque, crystallized prepolymer by treating it with a suitable solvent, or by heating it at a temperature higher than the glass transition temperature (Tg) of the amorphous prepolymer.

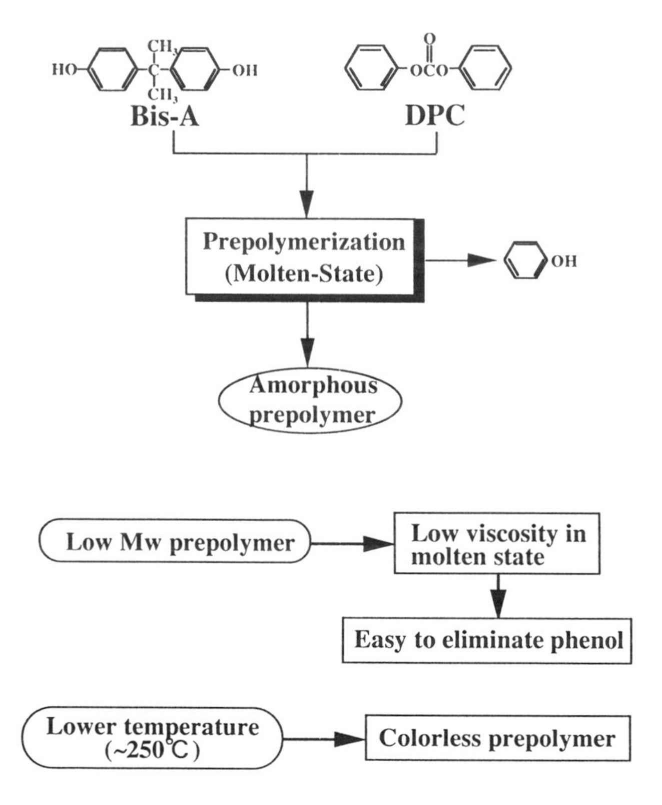

Figure 4. Prepolymerization step.

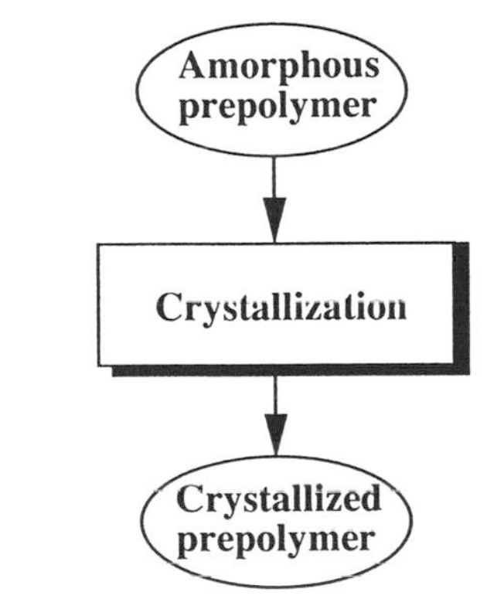

Crystallization of low molecular weight PC is very easy!!

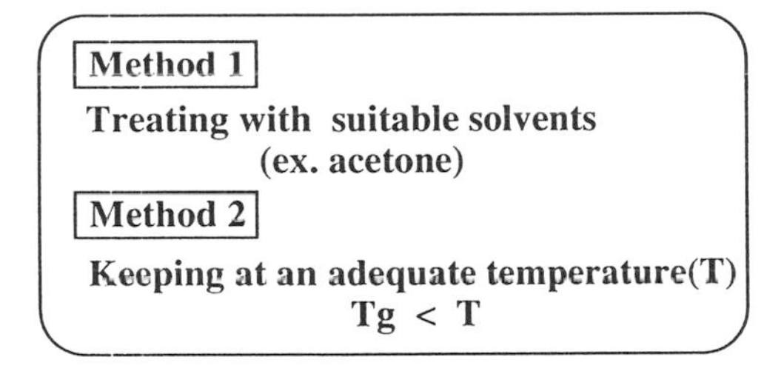

Figure 5. Crystallization step.

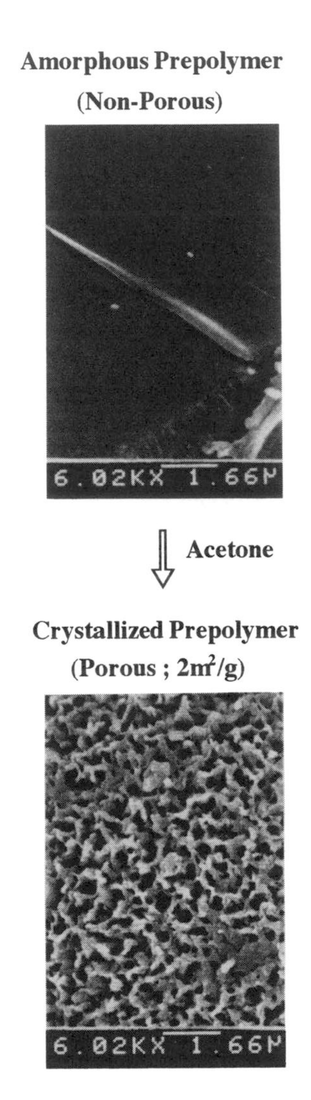

Figure 6. Scanning electron micrographs of prepolymers.

In this step, the prepolymer generally becomes porous at the same time it becomes crystallized. The solvent treatment of the amorphous prepolymer especially tends to give a crystallized prepolymer with a large surface area. In a typical example, the crystallized prepolymer obtained by contacting the amorphous, non-porous prepolymer with acetone, is a porous powder with 20% crystallinity and a melting point of 225°C. In this case, when the amorphous prepolymer is contacted with acetone, acetone permeates the prepolymer to cause lowering the Tg and raising the mobility of molecular chains. Subsequently the molecular chains rearrange to cause crystallization of prepolymer. As a progress of the crystallization, acetone is forced out from the prepolymer, and the crystallized prepolymer becomes porous. Figure 6 shows the scanning electron micrographs of a typical example of the crystallized prepolymer, which is highly porous, in contrast to the amorphous, non-porous prepolymer.

Solid-state Polymerization Step. The solid-state polymerization can be performed by heating the crystallized prepolymer at a temperature a little lower than its melting point under a flow of heated inert gas such as nitrogen, carbon dioxide, or lower hydrocarbons, or under reduced pressure (Figure 7). Since the surface area of the crystallized prepolymer is large, the rate of the solid-state polymerization is sufficiently fast. The polymerization, therefore, can be easily carried out at a lower temperature, for example, 210~220°C, in comparison to the melt process (280~310°C). Because the solid-state polymerization conditions are mild for the polycarbonate, the obtained polycarbonate is colorless and has several excellent properties.

Another excellent feature of the solid-state polymerization process of the polycarbonate is that the process has made it possible to produce a very high molecular weight polycarbonate with high purity. It is impossible to produce such an ultra high molecular weight polycarbonate by the conventional process. Figure 8 shows the molecular weight ranges of the polycarbonates achieved by the different processes. The phosgene process covers a range from disk grade to high viscosity grade, and the melt process covers an even smaller range, but the solid-state polymerization process covers a very wide range from lowest disk grade up to the ultra high molecular weight grade.

The solid-state polymerization can be carried out as either a batchwise processes or a continuous processes.

Benefits of Asahi's Solid-State Polymerization Process

Quality of the Polycarbonate by Solid-state Polymerization Process. Asahi's new polycarbonate process is not only environmentally benign, but is also able to produce polycarbonates with very high quality.

Asahi's polycarbonates are colorless with good transparency; other excellent features include, for example, heat stability and reworkability, in contrast to polycarbonates produced by the phosgene process. Asahi's polycarbonates are higher quality because they are free from impurities such as chlorinated compounds, which are difficult to remove from polycarbonates obtained from the phosgene process and which have a negative effect on polymer properties.

Furthermore, the ultra high molecular weight polycarbonate produced by the solid-state polymerization process also has excellent properties, which are missing in the conventionally available polycarbonates, such as excellent solvent resistance and excellent steam resistance. For example, after steam treatment (129°C, 20hrs),

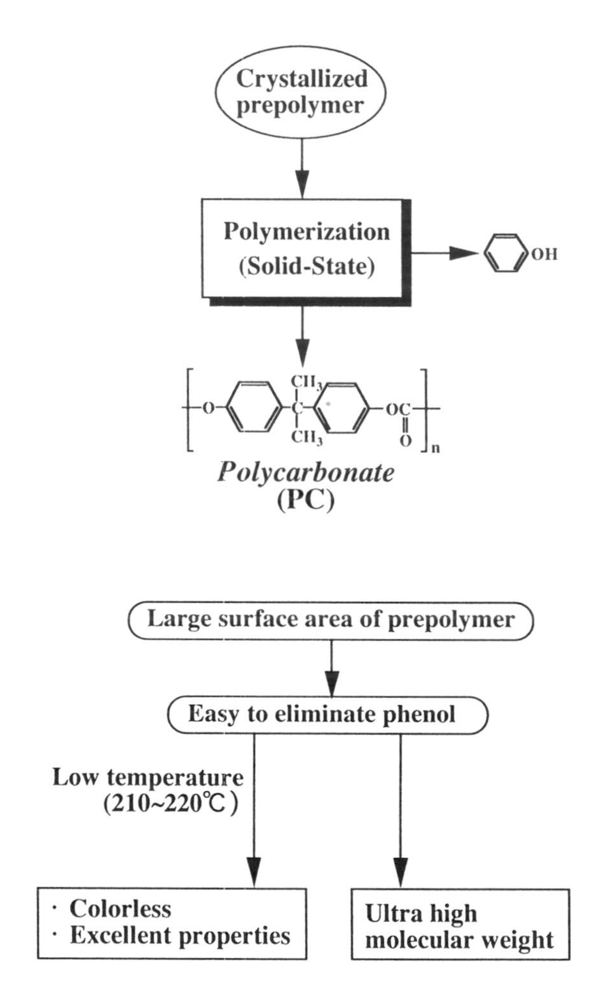

Figure 7. Solid-state polymerization step.

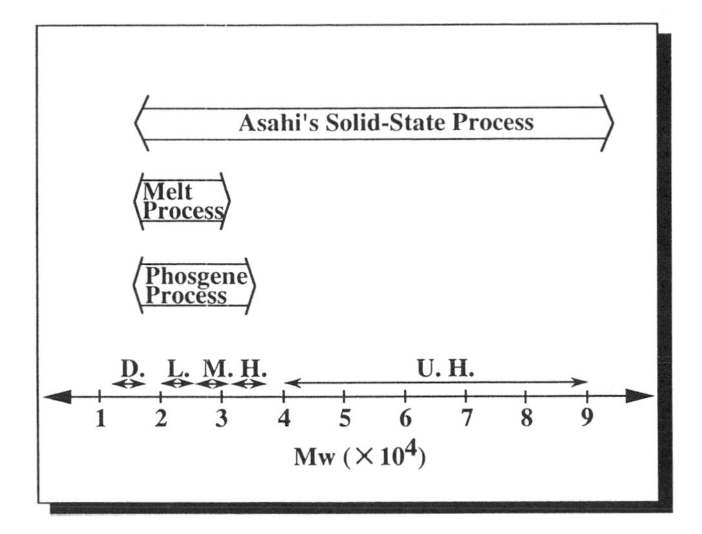

Figure 8. Comparison of the molecular weights achieved with the different processes. (D.) Disk grade; (L.) Low viscosity grade; (M.) Medium viscosity grade; (H.) High viscosity grade; (U.H.) Ultra high molecular weight grade.

the ultra high molecular weight polycarbonate kept its transparency, but the conventionally available polycarbonate lost its transparency.

The Production Costs of Asahi's Polycarbonates. Asahi's new process is able to produce polycarbonate of higher quality than that of the conventional phosgene process and, more importantly, the production cost of the new process has been estimated to be completely competitive with that of the phosgene process.

Conclusion

In this paper, the new polymerization technology for manufacturing polycarbonate is described. This new process contributes to pollution prevention, because it needs to use neither highly toxic phosgene nor methylene chloride, a probable human carcinogen.

In addition to the development of a polymer process which is environmentally benign, an innovative non-phosgene monomer process for producing diphenyl carbonate has been developed via dimethyl carbonate, using carbon dioxide as a raw material. This monomer process, which will be reported in other papers, has also been estimated to be completely competitive with other processes already commercialized or under development. Therefore, high-quality polycarbonates are now able to be produced from carbon dioxide and bisphenols.

These monomer and polymer processes are more environmentally benign than current processes for two reasons, namely the elimination of both phosgene and methylene chloride, and the utilization of carbon dioxide.

Literature Cited

1. (a) Fukuoka, S.; Chono, M.; Kohno, M. CHEMTECH. 1984, 670-676.
(b) Fukuoka, S.; Chono, M.; Kai,T.; Kohno, M. A New Route to Isocyanate Without Phosgene; Ashida,K.; Frisch, K.C., Ed.; International Progress in Urethanes; Technomic Publishing Co. Inc., Lancaster, PA.,1988, vol.5; pp 1-27
2. Schnell, H. Chemistry and Physics of Polycarbonates; Interscience: New York; 1964
3. Matsukane, M.; Tahara, S.; Kato, S. Polycarbonate Resin (in Japanese); Nikkan Kogyo Shinbun Publishing Co.,: Tokyo, 1969
4. Yasuda, T. Engineering Materials (in Japanese) 1989, 37 (8), 101-110.
5. Hosaka, N. Plastics (in Japanese) 1994, 45 (2), 35-39
6. However, only recently, GE Plastics Japan has announced the commercialization of the melt process.
7. Freitag, D; Grigo, U; Müller, P; Nouvertne, W Polycarbonates; Mark, H.; Bikales, N., Ed.; Encyclopedia of Polymer Science and Engineering; A Wiley-Interscience Publication: New York, 1988, vol. 11; 650-652.
8. Fukuoka, S.; Deguchi, R.; Tojo, M. US Patent 5,166,393, 1992
9. Fukuoka, S.; Tojo, M.; Kawamura, M. US Patent 5,210,268, 1993
10. Tojo, M.; Fukuoka, S. Japanese Patent Kokai 63-238043, 1988
11. Fukuoka, S.; Watanabe,T.; Dozono, T. US Patent 4,948,871, 1990
12. Fukawa, I.; Fukuoka, S.; Komiya, K.; Sasaki, Y. US Patent 5,204,377; 5,214,073, 1993
13. Flory, P. J. US Patent 2,172,374, 1939
14. Hagemeyer, H. J. US Patent 3,043,808, 1962

RECEIVED September 21, 1995

Chapter 3

Caprolactam via Ammoximation

G. Petrini[1,3], G. Leofanti[1,3], M. A. Mantegazza[1,3], and F. Pignataro[2]

[1]EniChem S.p.A.–Centro ricerche, Via S. Pietro 50, 20021 Bollate (Milan), Italy
[2]EniChem S.p.A., Via Taramelli 26, 20100 Milan, Italy

The new EniChem process for the production of cyclohexanone oxime, a chemical intermediate for Nylon 6, is described. The oxime synthesis is based on the cyclohexanone ammoximation using a new catalyst (Ti silicalite or TiS). The reaction operates in the liquid phase under mild conditions and proceeds in very high yield. The process overcomes the complexity and the disadvantages of the current technology, namely high inorganic salt byproduction and gaseous emission of SO_2 and NO_x. This paper compares the new technology with currently used methods and reports the reaction mechanism and the characteristics of the catalyst.

Over 90% of the worldwide installed production capacity of ε-caprolactam (CPL) is based on the synthesis of cyclohexanone oxime from cyclohexanone and a hydroxylamine derivative, followed by the Beckmann rearrangement to CPL with oleum (Figure 1). These current CPL processes differ mainly in the manufacture step of the hydroxylamine derivative. Nitrogen oxides from ammonia oxidation are reduced with SO_2 in the Raschig process and by catalytic hydrogenation in the BASF and DSM/Stamicarbon technologies (*1*). In all cases the product is a hydroxylammonium salt (sulfate or phosphate). The Raschig technology is characterized by SO_2 and NO_x emissions, the BASF and DSM/Stamicarbon processes produce only NO_x (*2*).

In the CPL production through the cyclohexanone oxime intermediate, the large byproduction of ammonium sulfate is a critical environmental and economic issue associated with the process, due in part to possible problems related to the disposal of ammonium sulfate.

Because of the problems associated with current CPL production, the focus of EniChem work has been the development of an "once-through" synthesis of cyclohexanone oxime that would satisfy the following requirements: 1) introduction

[3]Current address: EniChem/Base, CER, Via G. Fauser 4, 28100 Novara, Italy

0097–6156/96/0626–0033$12.00/0

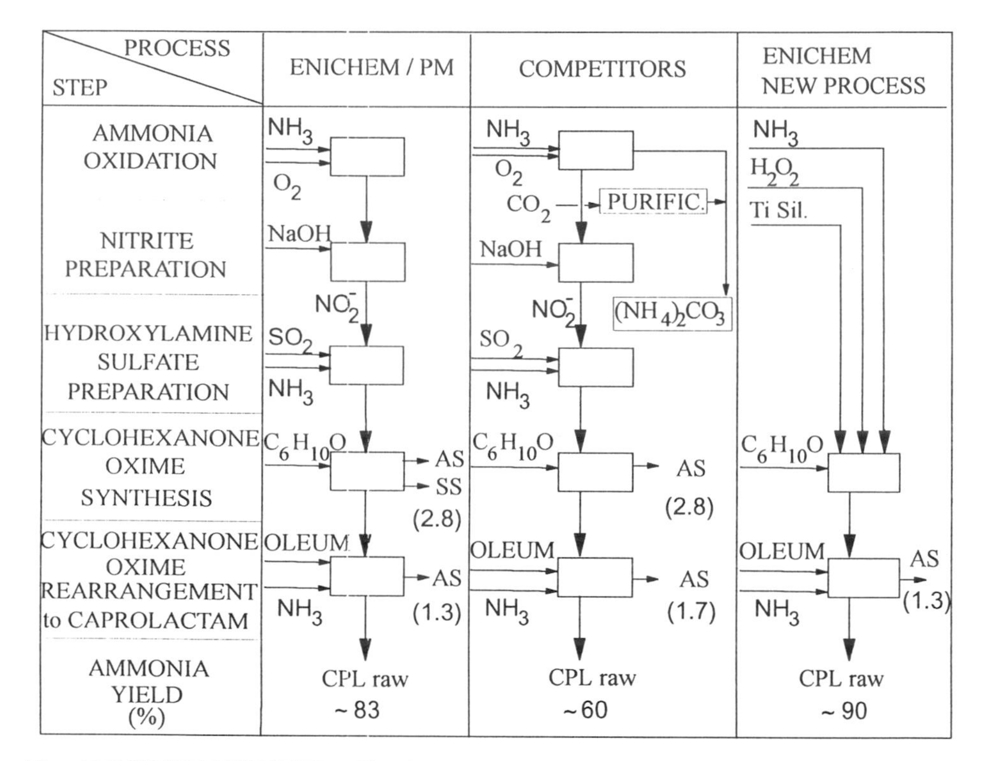

AS = AMMONIUM SULFATE ; SS = SODIUM SULFATE; ( ) = Kg/Kg of CAPROLACTAM

Figure 1. Comparison of Current CPL Technology Using Hydroxylamine Derivatives to the New EniChem Process.

of nitrogen into the cyclohexanone molecule by direct oxidation of ammonia to the correct oxidation state without over-oxidation to nitrogen oxides; 2) choice of an inexpensive oxygen source (air or low cost activated oxygen); 3) low or no byproduction of sulfate; 4) high yield and selectivity; and 5) simplification of the whole process and low investment costs. The cyclohexanone ammoximation reaction, reported previously (*1, 3*), satisfied most of these requirements but did not find industrial application initially because of the unsatisfactory yield.

Looking for a new catalyst, Roffia et al. (*4*) found that Ti containing systems were active and selective in catalyzing the ammoximation reaction when hydrogen peroxide was used as the oxidant. Ti silicalite, a zeolitic compound discovered by ENI researchers (*5*) and used in the hydroquinone process from phenol and hydrogen peroxide, was particularly efficient in the cyclohexanone ammoximation reaction. The reaction operates in liquid phase under mild conditions and proceeds in very high oxime yield. The development of the new EniChem ammoximation technology overcomes the disadvantages of the current processes, namely high inorganic salt byproduction and gaseous emission of SO_2 and NO_x.

Ammoximation Process

EniChem project was based on the following ammoximation reaction

$$C_6H_{10}O + NH_3 + 1/2\ O_2 \rightarrow C_6H_{10}NOH + H_2O \qquad (1)$$

previously studied both in the gas phase, Allied route (*1*), and in the liquid phase, Toagosei process (*3*).

The Allied process (*1*) involves gas phase ammoximation by oxygen on a silica gel catalyst. The low yield (45% on cyclohexanone), as well as the rapid fouling and low productivity of the catalyst, prevented valid applications of the method. The Toagosei process (*3*) involves ammoximation in the liquid phase and is catalyzed by phosphotungstic acid. The main disadvantages of this process are substantial loss of hydrogen peroxide (36%) and catalyst decomposition.

The reaction in liquid phase, however, appeared the most promising. As an oxidant, preference was given to hydrogen peroxide, a very selective oxidant under mild reaction conditions. An economic estimate shows that the cost of hydrogen peroxide for captive use makes the cyclohexanone ammoximation process competitive with the conventional processes.

The challenge was to discover a new catalyst that would be very selective in the cyclohexanone ammoximation reaction when using hydrogen peroxide as the oxidant. TiS, a zeolite containing Ti, was previously reported (*5*) as a very selective catalyst for epoxidation, hydroxylation and oxidation reactions using hydrogen peroxide. It has been found that the ammoximation reaction, when using TiS as the catalyst, proceeds in both high selectivity and yield (*6*).

The ammoximation reaction mechanism involves two steps (*7*). In the first step, ammonia is catalytically oxidized by hydrogen peroxide to hydroxylamine, the reaction intermediate:

$$NH_3 + H_2O_2 \rightarrow NH_2OH + H_2O \qquad (2)$$

The hydroxylamine yield can be as high as 70% on hydrogen peroxide (*8*). In the second step, the hydroxylamine reacts with cyclohexanone to give the cyclohexanone oxime:

$$C_6H_{10}O + NH_2OH \rightarrow C_6H_{10}NOH + H_2O \qquad (3)$$

The reaction was carried out as a semi-continuous process, at laboratory scale, by feeding the hydrogen peroxide to the stirred reaction mixture. The reaction conditions were as follows: solvent was *t*-butanol, catalyst concentration was 2 wt %, temperature was 80°C, NH_3/H_2O_2 molar ratio was 2.0 and reaction time was 5 hours (*6*). When providing a near stoichiometric H_2O_2/cyclohexanone molar ratio, the hydrogen peroxide loss was very small and was a result of the formation of some inorganic byproducts: ammonium nitrite, nitrate and nitrogen. Only traces of organic byproducts (deriving from side reactions of cyclohexanone in the basic medium) were formed.

The results obtained with different catalysts are reported in Table I. Another Ti containing catalyst, TiO_2 supported on amorphous silica, showed good catalytic properties but the best performances were obtained with TiS. When using amorphous silica or silicalite without Ti, the ammoximation occurred only to a negligible extent, as when no catalyst was used.

Table I. Cyclohexanone Ammoximation by Different Catalysts

Catalyst	Ti	$\frac{H_2O_2}{C_6H_{10}O}$	Cyclohexanone Conv.	Cyclohexanone Select.	H_2O_2 yield
	(wt %)	(molar ratio)	(mol. %)	(mol. %)	(mol. %)
None	-	1.07	53.7	0.6	0.3
SiO_2 amorphous	0	1.03	55.7	1.3	0.7
Silicalite	0	1.09	59.4	0.5	0.3
TiO_2/SiO_2	1.5	1.04	49.3	9.3	4.4
TiO_2/SiO_2*	9.8	1.06	66.8	85.9	54.0
Ti silicalite	1.5	1.05	99.9	98.2	93.2

*reaction time 1.5 hours
(Adapted from ref. 4).

The results obtained for the cyclohexanone ammoximation by TiS using different solvents are reported in Table II. *t*-Butanol proved to be the best solvent for the reaction, even if other solvents (toluene, benzene) were used with similar results.

Table II. Cyclohexanone Ammoximation by TiS in Different Solvents

Solvent	H_2O_2 / $C_6H_{10}O$	Cyclohexanone Conv.	Select.	H_2O_2 yield
	(molar ratio)	(mol. %)	(mol. %)	(mol. %)
Benzene	1.03	99.7	95.0	91.7
Toluene	1.07	99.8	97.0	90.0
t-amyl alcohol	0.86	94.5	95.6	94.0
t-butanol	1.05	99.9	98.2	93.2

(Adapted from ref. 4).

The results obtained for the ammoximation reaction run in *t*-butanol at different temperatures between 60 and 95 °C are shown in Table III. At 80 and 95°C the oxime selectivity and yield were similar and very high. At lower temperatures, both selectivity and yield decrease.

Table III. Cyclohexanone Ammoximation by TiS at Different Temperatures

Temperature	H_2O_2 / $C_6H_{10}O$	Cyclohexanone Conv.	Select.	H_2O_2 yield
(°C)	(molar ratio)	(mol. %)	(mol. %)	(mol. %)
60	0.93	81.5	87.0	76.4
70	0.97	90.2	96.4	89.4
80	0.83	80.1	98.8	95.0
95	0.88	83.0	99.9	94.0

(Adapted from ref. 4).

Starting from the semibatch results, a continuous microplant was built up for determining the stability and performance of the catalyst. More than 1600 hours of continuous operation of this microplant allowed both the specific consumption of the catalyst (acceptable from an economical point of view) and the stability of its performance to be determined. The conversion and selectivity were greater than 99% based on cyclohexanone and 90% based on hydrogen peroxide.

All unit operations necessary for the recovery of the cyclohexanone oxime, the rearrangement to CPL, and the subsequent product purification were also studied and optimized at the laboratory level.

Caprolactam Technologies Comparison. The ammoximation process simplifies a very complex part of the current CPL technology, namely the requirement of different hydroxylamine compounds for the cyclohexanone oxime synthesis. It is in these two steps, preparation of the hydroxylamine derivative and cyclohexanone oxime synthesis, that most of the undesired byproducts (NO_x, SO_2 and ammonium sulfate) are formed.

In Table IV the byproduction of ammonium sulfate in the ammoximation process is compared to other CPL processes. As shown in Table IV, 3/4 of the ammonium sulfate byproduction is eliminated in the EniChem process compared to the other processes. Finally, the data in Table V show that the ammoximation unit does not produce any consistent quantity of gaseous pollutants.

Table IV. Ammonium Sulfate Byproduction in CPL Processes

Process	Ammonium Sulfate Byproduction (Kg/kg of CPL)			Market Share (%)
	Oximation	Rearrangement	Total	
from cyclohexanone				
ENICHEM/RASCHIG	2.8	1.3	4.1	5
RASCHIG	2.8	1.6	4.4	45
BASF	1.0	1.6	2.6	21
DSM/STAMICARBON	0.0	1.6	1.6	20
ENICHEM Ammoximation	0.0	1.3	1.3	--
other processes	--	--	3-5	9

Table V. Nitrogen and Sulfur Derivatives Emissions of CPL Processes

PROCESS	SO_2	NO_x	NH_3	$NH_3(SO_x)$	NO_2^-/NO_3^-	OXIME
Raschig	2.65	3.3	.85	1.4 (2.64)	0.5	0.2
Ammoximation	----	----	----	---	1.0	----

[as Kg of NH_3 (SO_2)/1000 Kg of Oxime]

A comparison of the current technology and the new EniChem process is shown in Figure 1. The dramatic simplicity of the ammoximation step is evident.

Catalyst Characteristics and Reaction Mechanism

The TiS, discovered by Taramasso et al., belongs to the ZSM-5 pentasil family (*5*). A schematic representation is shown in Figure 2. The structure is formed by a proper linkage of $[SiO_4]$ units to give both straight and sinusoidal channels that are approximately 5.5 Å in width. In the computer generated scheme shown in Figure 2, only the straight channels are visible. The solid points represent Ti atoms.

Under vacuum conditions the most important spectroscopic features associated with Ti can be summarized as follows:

1) IR band (Figure 3a) at 960 cm^{-1} associated with a Si-O stretching mode in $[SiO_4]$ units perturbed by adjacent $[TiO_4]$ units and with a stretching mode of $[TiO_4]$ units (*9, 10, 11*);
2) Raman bands (Figure 3b) at 960 cm^{-1} (strong) and at 1127 cm^{-1} (weak) which can be assigned to two Raman active modes of perturbed $[SiO_4]$ units, or to two vibrations involving $[TiO_4]$ units embedded in the lattice (or to a mixture of the two modes) (*11*);
3) optical transition at 48000 cm^{-1} (Figure 4, spectrum 1) with ligand to metal charge transfer (LMCT) character in tetracoordinated and isolated Ti(IV) (*7, 9, 12*);
4) X-ray absorption (Figure 5) in TiK pre-edge region, with peak position, full width half maximum (FWHM) and intensity indicating that Ti is tetracoordinated and in a symmetry very close to a perfect tetrahedron (*13*).

The data point out that, in TiS, an isomorphous substitution of framework Si atoms by Ti atoms takes place, such that all Ti atoms are isolated and in tetrahedral coordination.

After the adsorption of water or ammonia, the following changes in the spectroscopic features have been observed:

1) a shift of the 960 cm^{-1} band (Figure 6) to 976 cm^{-1} (water) or 1050 cm^{-1} (ammonia) and the disappearance of the 1127 cm^{-1} band, showing the coordination of water and ammonia to Ti centers to give six-fold coordinated species (*11*);
2) a shift of Ti(IV) CT band to lower frequency (Figure 4, spectra 2, 3) due to Ti(IV) hexacoordinated complexes formed by ligand addition (*7*);
3) a disappearance of the pre-edge peak (Figure 5) replaced by a new weaker absorption characterized by very large FWHM, clearly indicating the formation of distorted octahedral species (*13*).

These results, together with those obtained by volumetric adsorption measurements, demonstrate that framework Ti atoms in TiS are able to coordinate with up to two ligands to reach their typical hexacoordinated status.

In Figure 4 the DRS UV-Vis spectra of TiS in the presence and in the absence of extra ligands are shown (*7*). The region of interest in these spectra is the area that represents the ligand to metal charge transfer (CT). In this

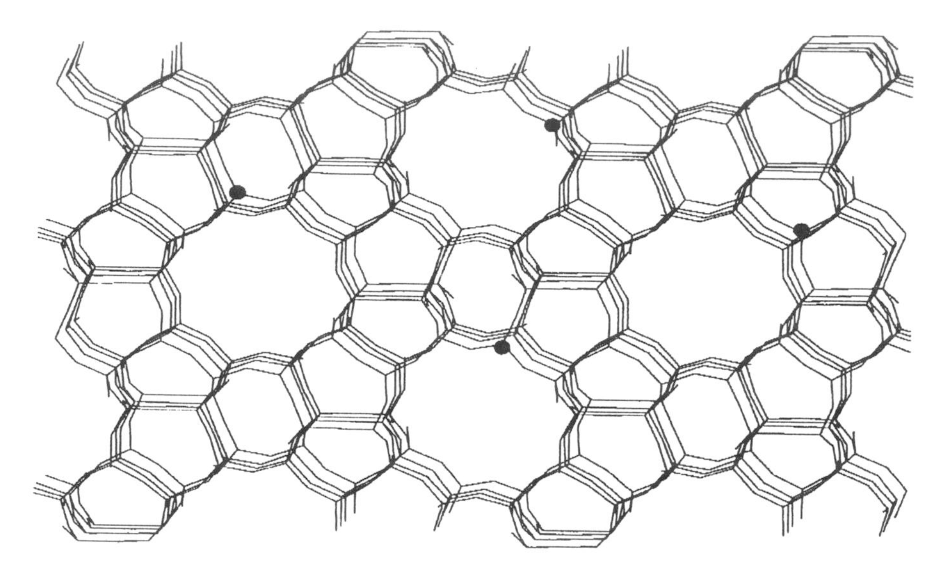

Figure 2. Schematic Representation of the Ti Silicalite Framework.

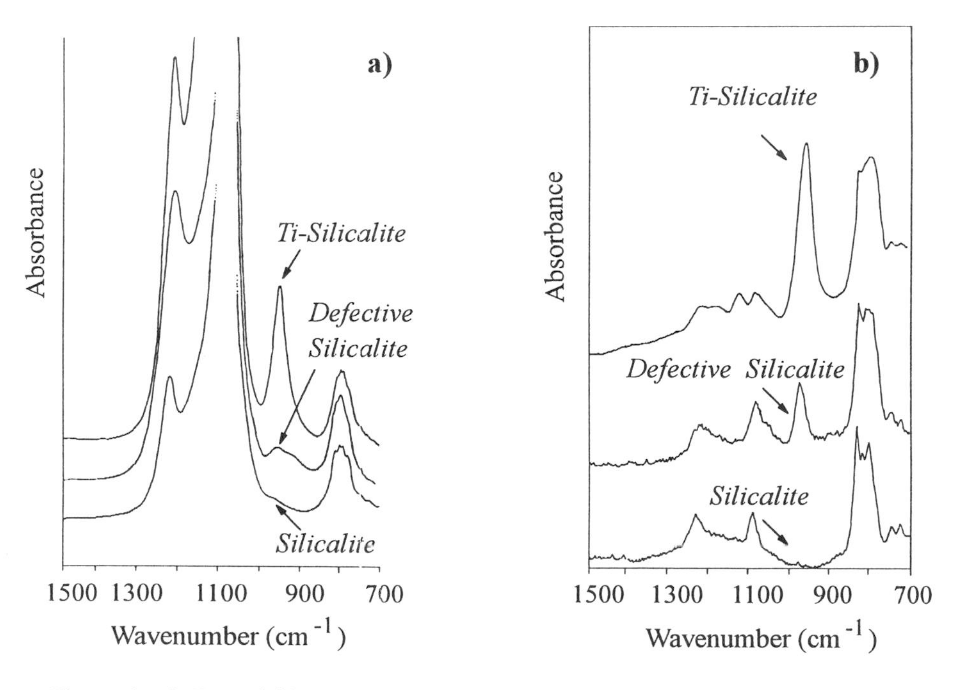

Figure 3. a) IR and b) Raman Spectra of Two Silicalite and One Ti Silicalite Samples. (Adapted from ref. 11).

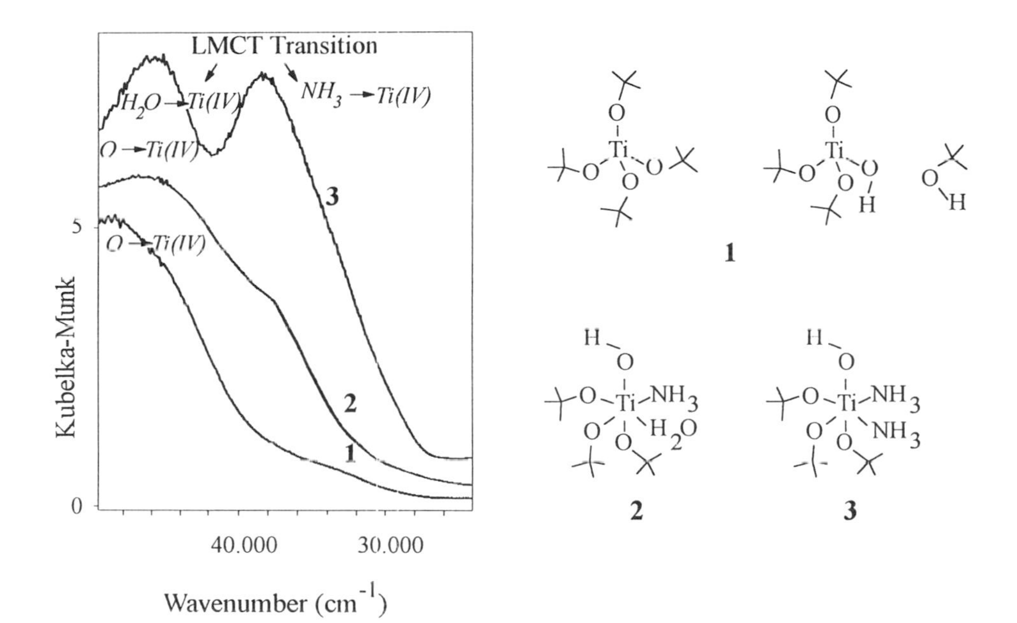

Figure 4. DRS UV-Vis Spectra of 1) Ti Silicalite, 2) Ti Silicalite in Presence of Water, 3) Ti Silicalite in Presence of Ammonia. (Adapted from ref. 7).

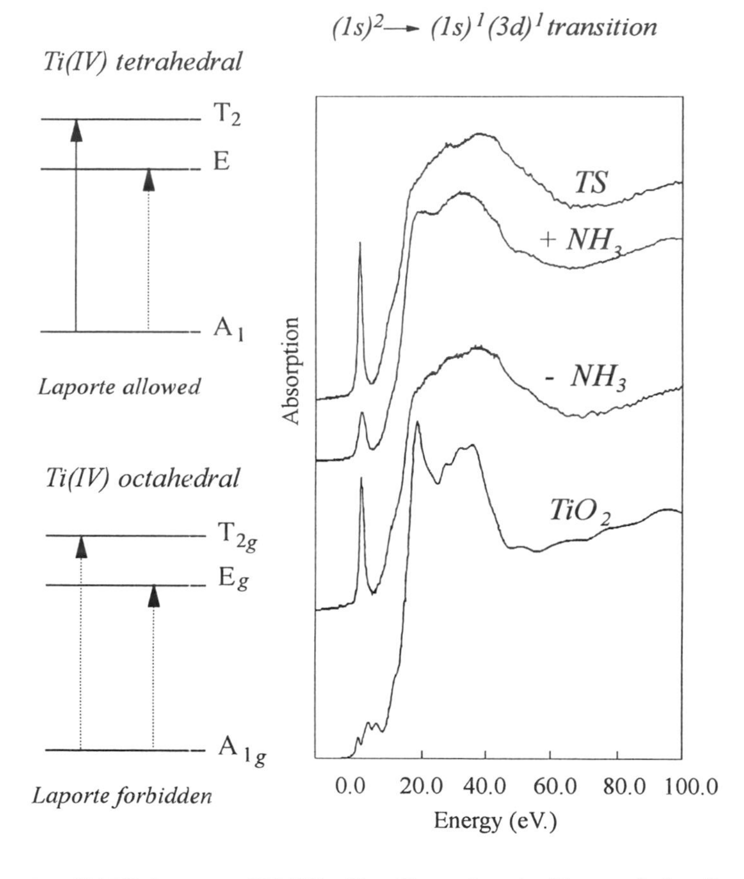

Figure 5. XAFS Spectra of Ti Silicalite. (Reproduced with permission from ref. 13. Copyright 1994 AIP (Woodbury).)

region, silicalite does not have any meaningful absorption. In spectrum 1 we observe a very weak absorption band at 30000 cm^{-1} and a very strong absorption band at 48000 cm^{-1}. Both absorptions represent the oxygen to Ti CT: the weak band, characteristic of an anatase-like phase (Ti is hexacoordinated and non isolated), indicates the presence of an impurity (about 0.03 wt%); the strong band is interpreted as an oxygen to Ti CT associated with isolated and tetracoordinated Ti centers as represented by structure 1 in Figure 4 (*12*).

The position of the 48000 cm^{-1} band depends both on coordination number and type of ligand. It is important to point out that in the DRS UV-Vis spectra, changes in the Ti coordination sphere result in spectral changes.

The modification induced by the interaction of Ti with extraligands, water and ammonia, is demonstrated in spectra 2 and 3 of Figure 4. Ti changes its coordination sphere from 4 to 6 after adding ammonia and water (structures 2 and 3). In fact, the two new peaks represent two new ligand to Ti CT transitions associated with the water and ammonia ligands (*7*).

The position and shape of the band at 48000 cm^{-1} and its complete sensitivity to the interaction with extraligands support the conclusion that Ti centers in TiS are isolated, tetracoordinated and available for chemical interactions.

Very interesting results were obtained on investigating the interaction of Ti centers with hydrogen peroxide. The DRS UV-Vis spectra of TiS in the presence and in the absence of water and hydrogen peroxide are reported in Figure 7. Compared to the DRS spectra of TiS (spectrum 1a, oxygen to metal CT) and TiS plus water (spectrum 2a, water to metal CT), the spectrum in the presence of hydrogen peroxide develops a new strong band in the region 22000-24000 cm^{-1}. The band was assigned to the CT from a hydroperoxo group to the Ti center (structure 3a). This interpretation is confirmed by spectrum *b*, where a very similar band appears in a peroxofluorotitanate compound (*14*).

The DRS UV-Vis spectra obtained with ammonia and hydrogen peroxide, i.e. two reagents in the ammoximation reaction, could also be very useful for investigating the elementary steps occurring at the Ti centers of this TiS-catalyzed reaction.

The DRS UV-Vis spectra reported in Figure 8 provide an explanation in the role of Ti center in the hydroxylamine synthesis as well as in the reaction sequence shown in the figure: TiS (spectrum 1) reacts with hydrogen peroxide in aqueous solution generating a hydroperoxo species (spectrum 2, complex 2); the addition of ammonia produces a mixed ammonia-hydroperoxo species (spectrum 3, complex 3). In time, ammonia is oxidized to hydroxylamine restoring the Ti center. Spectrum 4 shows the final step of the catalytic reaction (*7*).

Catalyst Deactivation. The behavior of the catalyst and its transformation in the reaction medium has been extensively investigated (*16*). Due to the characteristics of amorphous and crystalline silica (*15*), TiS interacts with the alkaline reaction medium and slightly dissolves. The TiS dissolution is a slow process depending on the reaction conditions and is controlled in the

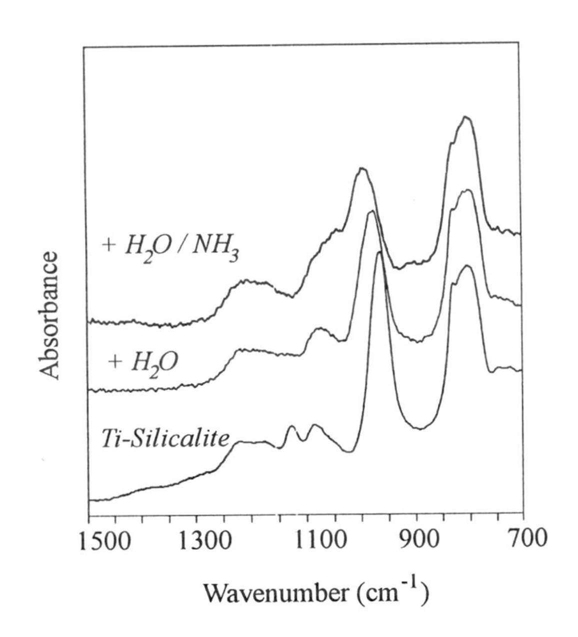

Figure 6. Raman Spectra of Ti Silicalite, Ti Silicalite in Presence of Water and in Presence of Ammonia. (Reproduced with permission from ref. 11. Copyright 1993 Royal Society of Chemistry).

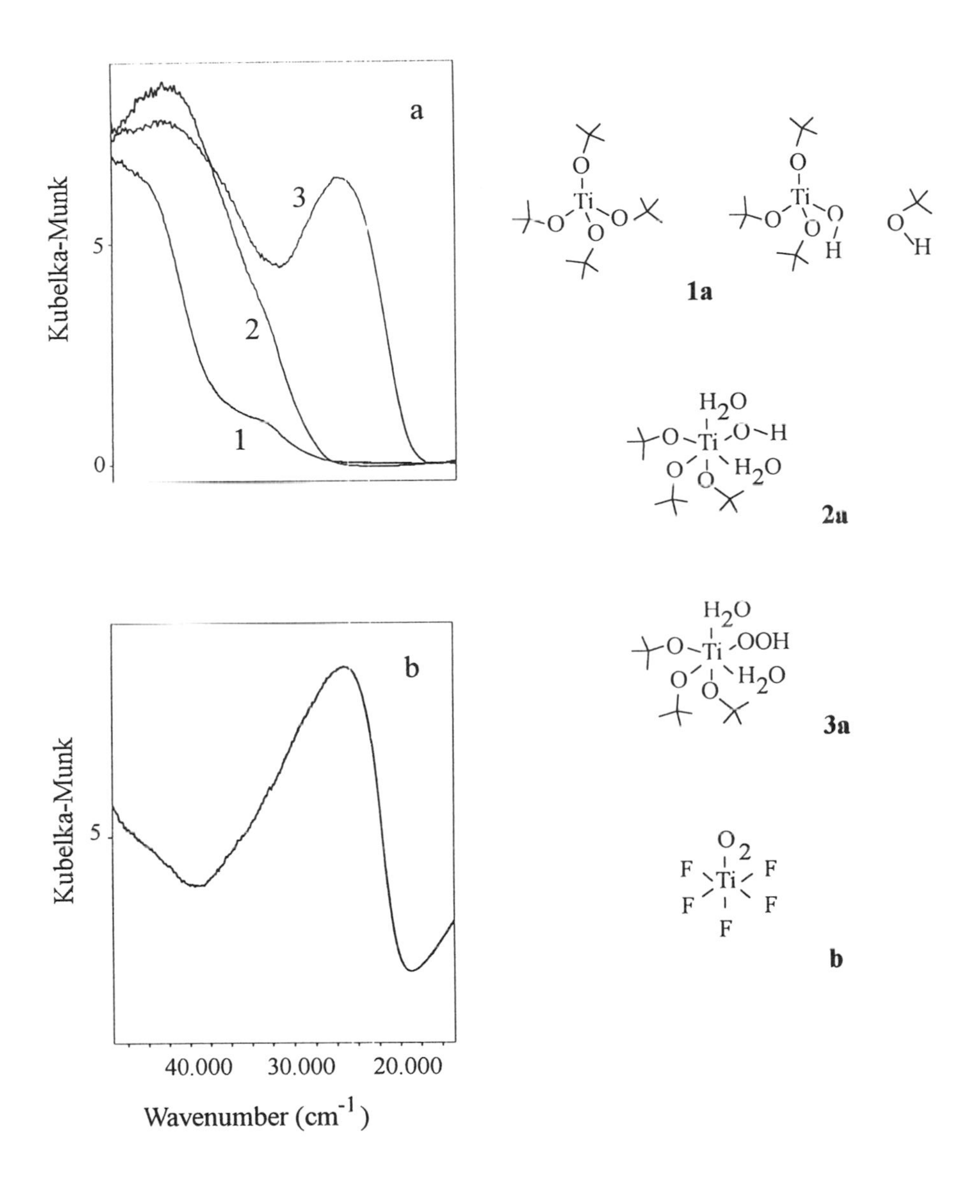

Figure 7. DRS UV-Vis Spectra of: a) Ti Silicalite in Absence or in Presence of Extraligands (H_2O, H_2O_2), b) $(NH_4)_3(Ti[O_2]F_5)$-H_2O. (Reproduced with permission from ref. 14. Copyright 1992 J. C. Baltzer AG).

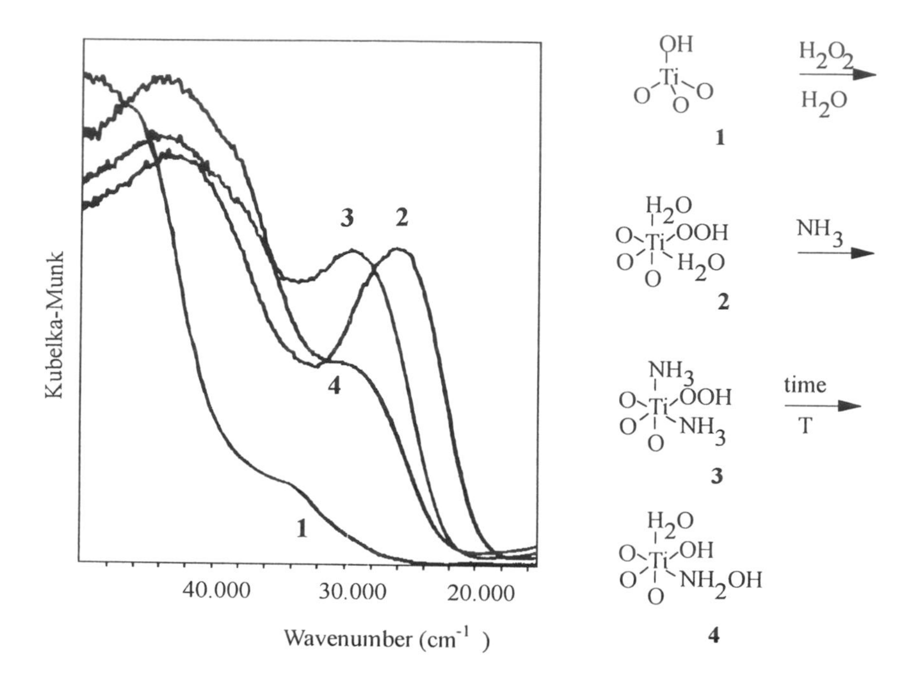

Figure 8. Hydroxylamine Formation on Ti Silicalite. (Adapted from ref. 7).

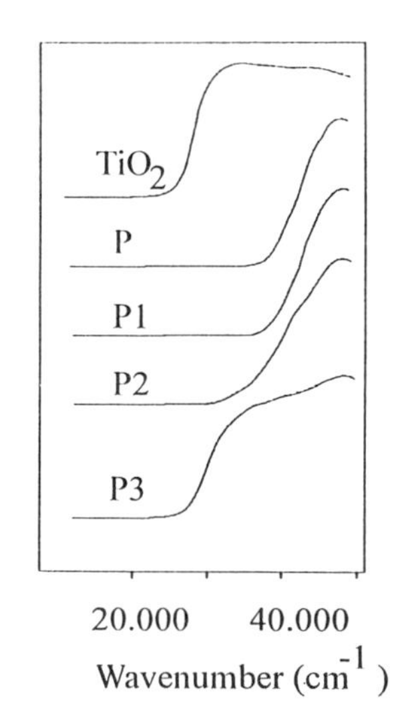

Figure 9. UV-Vis spectra of three aged Ti Silicalite samples. (Reproduced with permission from ref. 16. Copyright 1991 Elsevier Science B.V.).

ammoximation process. As a consequence of dissolution, Ti is continuously extracted from the zeolitic framework and accumulates on the residual catalyst as a Ti rich species, resulting in a less active solid (*16*).

The aging phenomena is illustrated by the DRS UV-Vis spectra of fresh catalyst (P) and catalyst samples (P1, P2 and P3) taken at different intervals during a continuous run of several hundred hours (Figure 9). The spectrum of TiO_2 (anatase) is also reported as a reference. The progressive formation of an anatase-like species (absorption at 42000 cm^{-1}) is evident from these spectra.

Conclusions

A new, highly innovative process for the production of cyclohexanone oxime, an intermediate in CPL production, has been described. The ammoximation reaction proceeds via a hydroxylamine intermediate, which is formed by the oxidation of ammonia with hydrogen peroxide and catalyzed by TiS.

The activity of Ti centers in the catalyst is due to: 1) the isolated and t-etracoordinated structure of the Ti atoms inserted in the silicalite framework; 2) their capability to coordinate with two new ligands (water, ammonia); and 3) their reactivity towards hydrogen peroxide to form hydroperoxo species. The catalyst deactivation is related to: 1) a slow dissolution of the framework with accumulation of Ti on the external surface of the remaining solid and 2) a partial extraction of Ti from the framework.

The main advantages of the new technology are: 1) simplification of the entire CL production process; 2) significant reduction of ammonium sulfate byproduction; and 3) total elimination of NO_x and SO_2 emissions.

The results reported have very high industrial applicability. The new process can be easily applied to all current CPL production technologies that are based on cyclohexanone oxime (90% of worldwide caprolactam production). The new technology is also suitable for the production of many other oximes.

An industrial demonstrative unit of 12 Kt/y CPL has been built up in the Porto Marghera site and is now at the start up phase. After converting the whole industrial line at the Porto Marghera site (135 Kt/y CPL) to the new technology, EniChem will be operating the most efficient, economical and environmentally benign caprolactam process.

Literature Cited

1. Armor, J. N. In *Catalysis of Organic Reactions;* Kosak, J. R., Ed.; Dekker: New York, 1984; 409.
2. *Kirk-Othmer Encyclopedia of Chemical Technology,* Kroschwitz J. I., Ed.; Wiley J. & Sons: New York, 1992, Vol. 4; 827-839.
3. Tsuda, S. *Chem. Econ. Eng. Rew.* **1970**, 39.
4. Roffia, P.; Leofanti, G.; Cesana, A.; Mantegazza, M.; Padovan, M.; Petrini, G.; Tonti, S.; Gervasutti P. In *New Developments in Selective Oxidation;* Centi, G. et al., Eds.; Elsevier: Amsterdam, 1990; 43.
5. Taramasso, M.; Perego, G.; Notari, B. U.S. Patent 4.410.501, 1983.

6. Roffia, P.; Padovan, M.; Leofanti, G.; Mantegazza, M.A.; De Alberti, G.; Tauszik, G.R. U.S. Patent 4.794.198, 1988.
7. Zecchina, A.; Spoto, G.; Bordiga, S.; Geobaldo, F.; Petrini, G.; Leofanti, G.; Padovan, M.; Mantegazza, M. A.; Roffia, P. In *New Frontiers in Catalysis;* Guzci L., et al., Eds; Akadémiai Kiadó: Budapest, 1993, 719.
8. Mantegazza, M. A.; Leofanti, G.; Petrini, G.; Padovan, M.; Zecchina, A.; Bordiga, S. In *New Developments in Selective Oxidation II;* Cortés Corberán, V. et al., Eds.; Elsevier: Amsterdam, 1994; 541.
9. Zecchina, A.; Spoto, G.; Bordiga, S.; Ferrero, A.; Petrini, G.; Leofanti, G.; Padovan, M. In *Zeolite Chemistry and Catalysis;* Jacobs, P. A. et al., Eds.; Elsevier: Amsterdam, 1991; 251.
10. Zecchina, A.; Spoto, G.; Bordiga, S.; Padovan, M.; Leofanti, G.; Petrini, G. In *Catalysis and Adsorption by Zeolites*, Ohlman G., et al., Eds.; Elsevier: Amsterdam, 1992; 671.
11. Scarano, D.; Zecchina, A.; Bordiga, S.; Geobaldo, F.; Spoto, G.; Petrini, G.; Leofanti, G.; Padovan, M.; Tozzola, G.; *J. Chem. Soc., Farad Transaction* 1993, 89 (22), 4123.
12. Boccuti, M. R.; Rao, K. M.; Zecchina, A.; Leofanti, G.; Petrini, G. *Structure and Reactivity of Surfaces*; Elsevier: Amsterdam, 1989, 133.
13. Bordiga, S.; Coluccia, S.; Lamberti, C.; Marchese, L.; Zecchina, A.; Boscherini, F.; Buffa, F.; Genoni, F.; Leofanti, G.; Petrini, G.; Vlaic, G. *J. Phys. Chem.* **1994**,. *98*, 4125.
14. Geobaldo, F.; Bordiga, S.; Zecchina, A.; Giamello, E.; Leofanti, G.; Petrini, G.; *Catal. Lett. 16*. 1992, 109.
15. Pascal, R. In *Nouveau Traité de Chimie Minerale*; Masson et C.ie, Eds, Paris, 1965, Vol. 8, 432.
16. Petrini, G.; Cesana, A.; De Alberti, G.; Genoni, F.; Leofanti, G.; Padovan, M.; Paparatto, G.; Roffia, P. In *Catalyst Deactivation 1991;* Bartholomew G. H., et al., Eds.; Elsevier: Amsterdam, 1991, 761.

RECEIVED September 21, 1995

Chapter 4

Generation of Organic Isocyanates from Amines, Carbon Dioxide, and Electrophilic Dehydrating Agents

Use of *o*-Sulfobenzoic Acid Anhydride

William D. McGhee, Mark Paster, Dennis Riley, Ken Ruettimann, John Solodar, and Thomas Waldman

Monsanto Corporate Research, 800 North Lindbergh Boulevard, St. Louis, MO 63167

Isocyanates are produced on the billions of pounds scale annually and are generated almost exclusively via phosgenation technology. The reaction of an amine with phosgene has several drawbacks, including the high toxicity of phosgene and the generation of HCl. The use of carbon dioxide as a phosgene replacement has been explored, and conditions for the highly selective synthesis of isocyanates from carbamate anions (generated from amines and carbon dioxide) have been found using various "dehydrating agents". One of the goals is not only to eliminate the use of phosgene in the production of isocyanates but also to eliminate salt waste. The initial results will be presented along these lines by the use of recyclable dehydrating agents.

In previous accounts the use of electrophilic dehydrating agents in the direct generation of isocyanates from amines and carbon dioxide has been discussed, equations 1-2 (*1-3*).

$$RNH_2 \ + \ CO_2 \ + \ Base \ \text{----------------->} \ RNHCO_2^{-}\,{}^{+}HBase \qquad (1)$$

$$RNHCO_2^{-}\,{}^{+}HBase \ + \ \text{Dehydrating agent} \ \text{-------->} \ RNCO \ + \ \text{salts} \qquad (2)$$

This chemistry was shown to proceed with the highest selectivities and yields using phosphorus containing electrophiles ($POCl_3$, PCl_3 and P_4O_{10}). Although this provided a relatively low cost, mild route to new materials, a large amount of salt waste was generated. Subsequently the use of non-halide and potentially low to no waste reagents capable of achieving the same high selectivities and yields of isocyanates (*4*) has been explored.

0097–6156/96/0626–0049$12.00/0

Introduction

From the outset of this research the use of organic anhydrides as dehydrating agents (i.e., acetic anhydride) was investigated with the idea of a "no waste" process for the generation of isocyanates via "dehydration" of carbamates (see Scheme I).

Scheme I

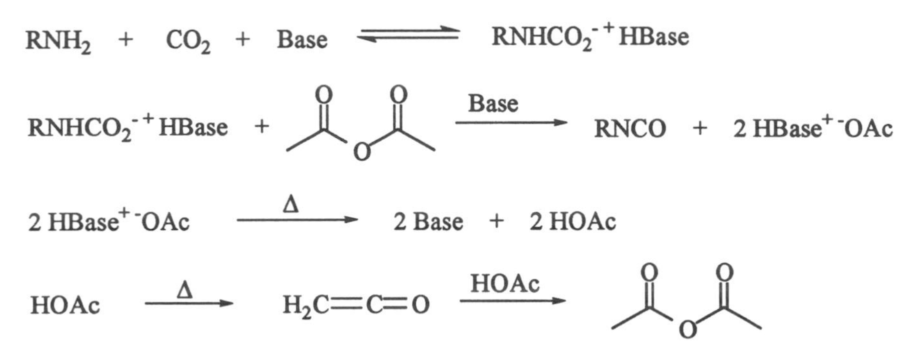

Although isocyanates were generated, the use of acetic anhydride, unfortunately, showed selectivities which were not of practical use (competitive generation of amide, see equation 3).

$$RNHCO_2^{-}\,{}^{+}HBase + (CH_3CO)_2O \xrightarrow{-CO_2} CH_3C(O)NHR + HBase^{+}\,{}^{-}OAc \quad (3)$$

Previous results clearly suggested that the leaving group ability on the anhydride plays an important role in the generation of isocyanates (see Table I).

Table I. "Dehydration" of Amines to Isocyanates Using Organic Anhydrides

RNH_2	Anhydride	Base	%NCO
octyl-NH_2	acetic	CyTEG	40
octyl-NH_2	acetic	NEt_3	2
octyl-NH_2	benzoic	CyTEG	65
octyl-NH_2	benzoic	NEt_3	22
octyl-NH_2	trifluoroacetic	CyTEG	95
octyl-NH_2	trifluoroacetic	NEt_3	35

All reactions were run in acetonitrile under 80 psig carbon dioxide at 0°C (CyTEG = N-cyclohexyl-N',N',N",N"-tetraethylguanidine).

The ability of an acetate group to act as a leaving group has been well established to be relatively poor in comparison to the excellent leaving group ability of the alkyl (aryl) sulfonates. Use of a sulfonate leaving group in the anhydrides gave excellent yields and selectivities of isocyanates (i.e., benzenesulfonic acid anhydride gave essentially quantitative yields of isocyanates under mild conditions), equation 4.

$$RNHCO_2^{-}\,{}^{+}HBase \;+\; C_6H_5{-}SO_2{-}O{-}SO_2{-}C_6H_5 \xrightarrow{\text{Base}} \qquad (4)$$

$$RNCO \;+\; 2\,C_6H_5{-}SO_3^{-}\,{}^{+}HBase$$

Unlike acetic anhydride, benzene sulfonic acid anhydride can not be generated from the parent acid via a thermal dehydration. This regeneration of anhydride is a critical step in a waste free process. Therefore, a dehydrating agent capable of high selectivity toward isocyanate generation and one which showed the ability toward thermal regeneration was sought.

Results and Discussion

Based on its availability and structural nature, investigation of the "dehydrating" power of *o*-sulfobenzoic acid anhydride (I) was carried out. The parent acid may be readily available as an intermediate in the synthesis of saccharin (*5*). The cyclic nature of this anhydride along with its half sulfonic/ half benzoic composition hinted to us that it may prove to be a suitable compound for the present isocyanate technology.

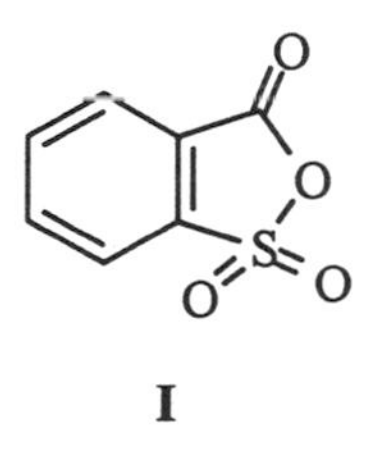

I

The results of the use of this anhydride in model reactions is summarized in Table II below. This reagent gave rise to high yields and selectivities of isocyanates under mild conditions.

Table II. Conversion of Amines to Isocyanates Using *o*-Sulfobenzoic Acid Anhydride

RNH_2	%NCO
n-octyl-NH_2	95
cyclohexyl-NH_2	100
triaminononane	83-92
HMD	70-80

All reactions run in acetonitrile using an excess of triethylamine as base under 80 psig carbon dioxide pressure at -10°C (triaminononane = 4-aminomethyl-1,8-octanediamine, HMD = 1,6-diaminohexane).

Ideally, the reaction proceeds as shown in Scheme II depicted below.

Scheme II

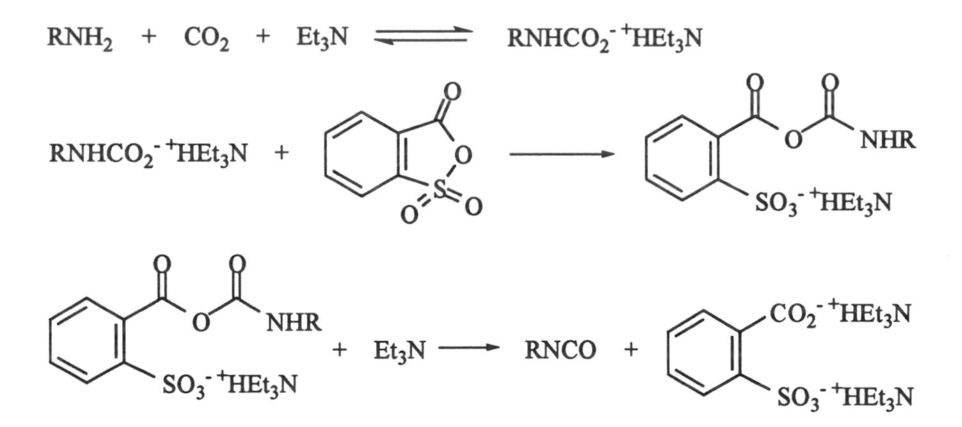

This reaction is thought to proceed via nucleophilic attack of the carbamate anion at the anhydride giving rise to a mixed anhydride which then undergoes base induced elimination to the corresponding isocyanate. The assumption that the intermediate is that resulting from nucleophilic attack at the carboxylic carbonyl rather than the sulfonyl site is based on literature precedence (*6*).

Extension of the use of this reagent to the synthesis of TTI (triaminononane tris-isocyanate) was investigated.

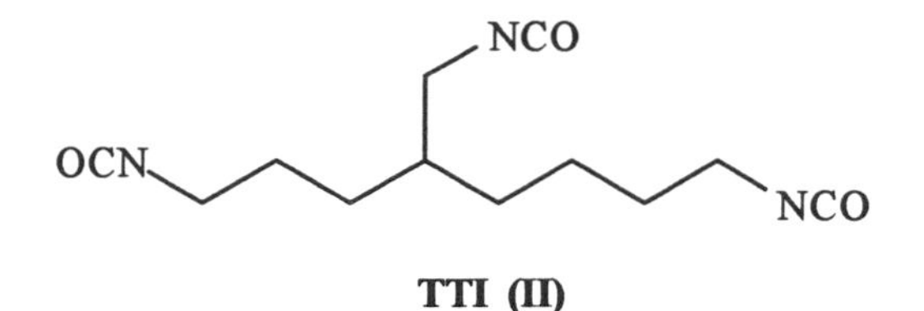

TTI (II)

The results for the reaction of triaminononane - tris carbamate with *o*-sulfobenzoic acid anhydride gave high yields of TTI (TAN triisocyanate)(*7-8*). This isocyanate has been evaluated as a low viscosity, low vapor pressure crosslinking agent in coating applications (*9*).

In the synthesis of this isocyanate using the above mentioned "dehydration" technology it was observed that an increase in isocyanate yield with an increase of one to two equivalents of anhydride (per reactive nitrogen) was found which was somewhat surprising. Insight into the cause of this result was gained by identifying the by-product of the reaction. Interestingly, it is the linear anhydride (III) which is the result of this dehydration reaction and not the sulfonate/benzoate as expected.

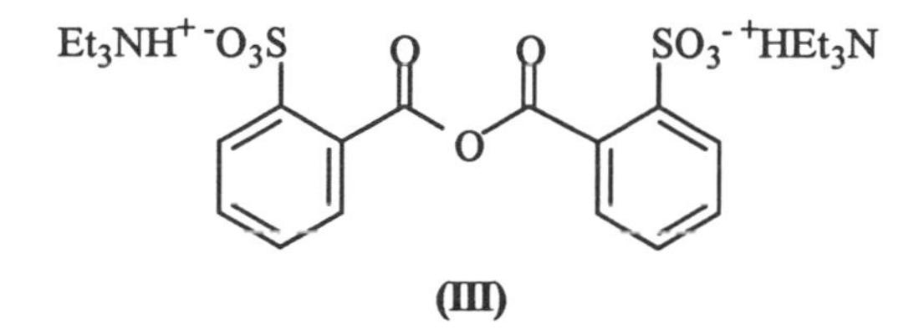

(III)

A reasonable explanation for the formation of the substituted benzoic acid anhydride from this reaction lies in competitive nucleophilic attack of generated carboxylate on the *o*-sulfobenzoic acid anhydride as shown below (*10*) in equation 5.

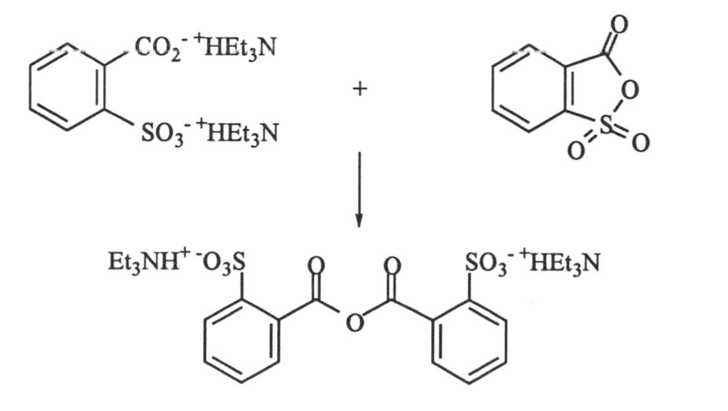

Independent confirmation of the plausibility of this reaction was demonstrated by the addition of an equal molar amount of the dianion of *o*-sulfobenzoic acid to the anhydride giving the same linear benzoic anhydride. This result points out that attack of the carboxylate is competitive with carbamate attack on the cyclic anhydride. The generation of linear anhydride does not appear to affect the production of isocyanate.

Thermal Reaction of Linear Anhydride. The most desirable and simplest scenario for regeneration of *o*-sulfobenzoic acid anhydride is the thermal cracking of the salt obtained, followed by thermal dehydration to the anhydride. Attempts to thermally effect this conversion have not been successful. Heating the linear anhydride in the solid state under a nitrogen purge generates one equivalent of *o*-sulfobenzoic acid anhydride and one equivalent of the sulfonate/benzoate as shown in equation 6. This is the reverse reaction of that observed for the formation of anhydride.

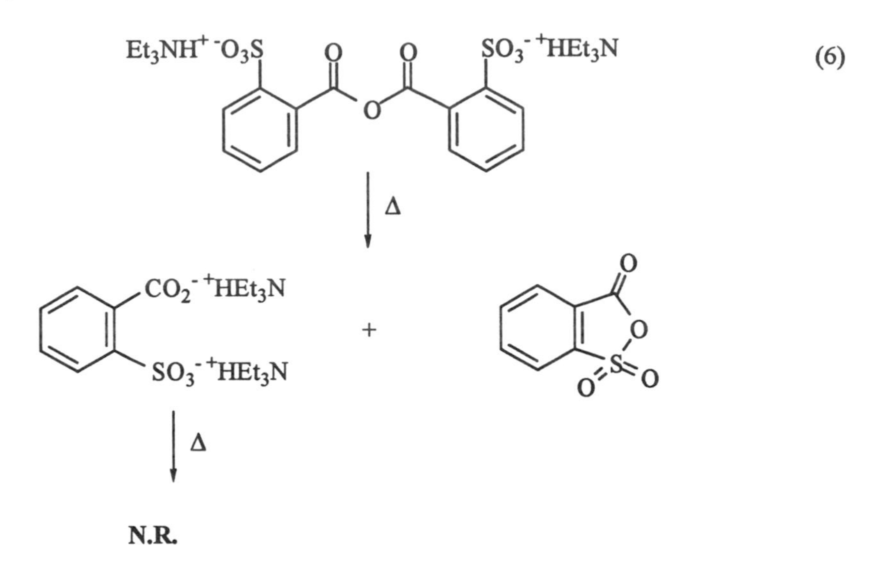

(6)

The anhydride sublimes out of the reaction leaving a heavy oil which was identified by IR. This result is consistent with the thermal gravimetric results obtained from the same linear anhydride. Continued heating of the remaining salt does not lead to any productive chemistry.

Conversion of Linear Anhydride to *o*-Sulfobenzoic Acid Anhydride. Critical to the successful use of *o*-sulfobenzoic acid anhydride as a dehydrating agent for the production of isocyanates is the ability to recycle the anhydride. Since direct conversion from the isolated by-product appears to be unlikely, another pathway must be demonstrated. A reasonable route is shown below in Scheme III, and progress in each of the steps shown has been made.

Scheme III

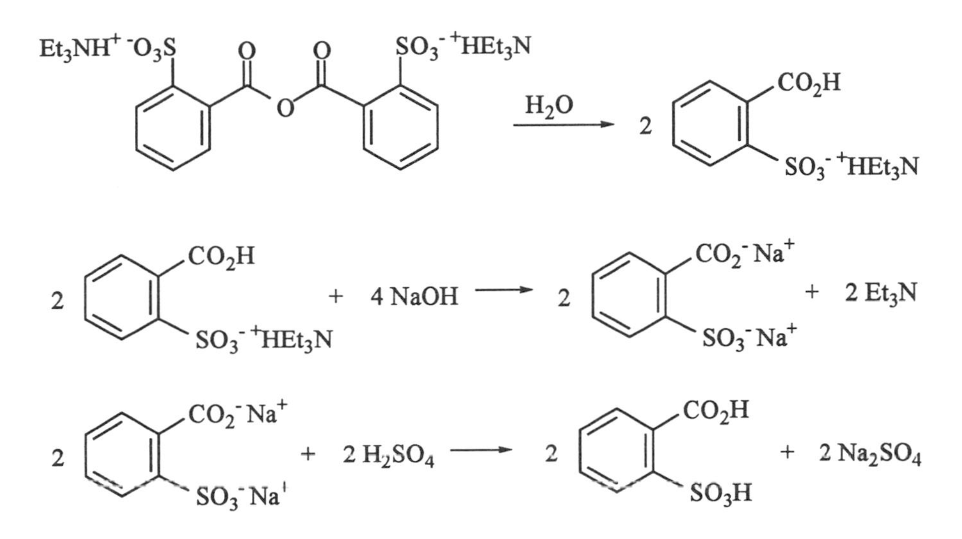

This route involves hydrolysis of the linear anhydride into the mono-triethylammonium salt followed by neutralization with caustic soda (or calcium hydroxide) to liberate the triethylamine giving the sodium salt of *o*-sulfobenzoic acid. This salt is then converted to the free acid via protonation by sulfuric acid (or on an ion exchange column, regeneration of the ion exchange resin involves addition of a strong acid such as sulfuric acid). The free acid is then dehydrated thermally (driven by a physical removal of generated water) to *o*-sulfobenzoic acid anhydride (*11*) as shown in equation 7.

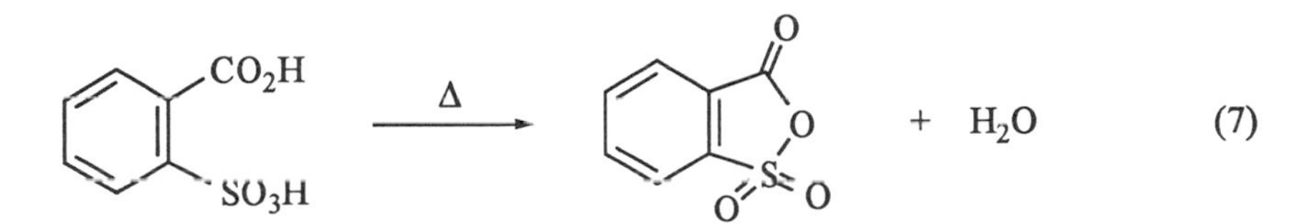

The above results show that a recyclable dehydrating agent for the high conversion of amines to isocyanates using carbon dioxide can be used; however, the use of sodium hydroxide (or calcium hydroxide) in the recycle of anhydride and for the recovery of triethylamine gives rise to salt by-product (sodium or calcium sulfate). To circumvent this salt generation one can envision a process that utilizes a "salt splitting" step (electrohydrolysis) (*12*). Use of this technology would create a process whereby we could generate isocyanates from amines and carbon dioxide with water as the sole by-product (see Scheme IV below).

Scheme IV

RNHCO2- +HEt3N + Et3N
RNCO
2
Et3NH+ -O3S
SO3- +HEt3N
H2O
Electrohydrolysis
CO2H
2
Heat
SO3H
2 Et3N

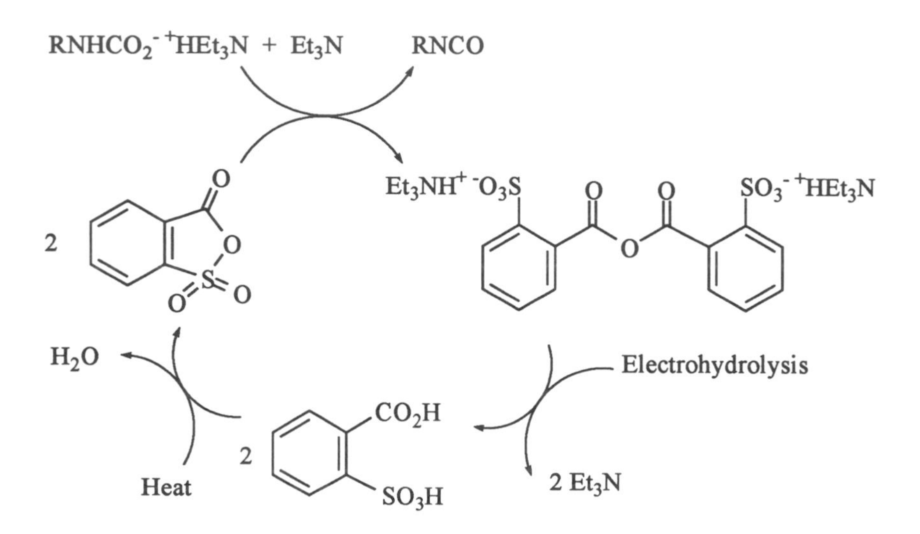

***In situ* trapping of isocyanate.** The mild conditions of the "dehydration" reaction using *o*-sulfobenzoic acid anhydride provided an opportunity to look at generating isocyanates *in situ* followed by immediate conversion of the isocyanate into urethane materials. This method may prove to be valuable for the generation of materials based on toxic, volatile isocyanates (i.e. methyl isocyanate). An example is shown below in Scheme V(*13*) which shows the formation of 1-naphthyl N-methylcarbamate (common insecticide).

Scheme V

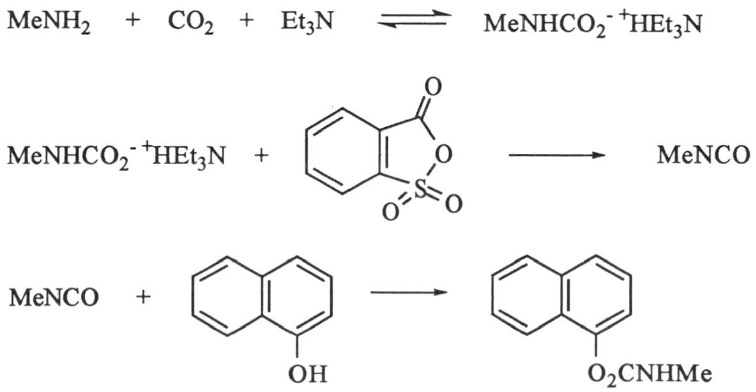

MeNH2 + CO2 + Et3N
MeNHCO2- +HEt3N
MeNHCO2- +HEt3N +
MeNCO
MeNCO +
OH
O2CNHMe

Experimental

Hexamethylene diisocyanate. Into a 3-neck round bottomed flask was charged 2.0 g (17 mmol) hexamethylenediamine, 15 mL triethyl amine (109 mmol), 0.262 g (1.7 mmol) biphenyl as G.C. internal standard and 100 mL acetonitrile. The three-neck flask was fitted with an overhead stirrer, a CO_2 gas inlet, a thermocouple and a dryice condenser. Into the dryice condenser was added a dryice/*m*-xylene slush bath (-48°C). Carbon dioxide was added subsurface to the reaction mixture at room temperature giving rise to a heterogeneous solution. Over 90 min. the reaction mixture was cooled to 0°C and then to -20°C (*o*-xylene/dryice bath). Once equilibration was established at -20°C *o*-sulfobenzoic acid anhydride (12.5 g, 68 mmol) was added as a solid slowly over a 30 min period. Aliquots were taken periodically and were quenched by the addition of toluene/aq. HCl followed by analysis by G.C. (78%).

1-Naphthyl N-methylcarbamate. Into a Fischer-Porter bottle was charged 2 g methyl ammonium methyl carbamate (19 mmol - 38 mmol methylamine equivalent; generated from methyl amine and carbon dioxide), 16 g triethylamine (158 mmol) and 75 mL CH_3CN. This was attached to a pressure head and 60 psig carbon dioxide was added above the reaction mixture giving rise to an exothermic reaction. Into a second Fischer-Porter bottle was added 13.9 g *o*-sulfobenzoic acid anhydride in 40 mL CH_3CN. After one hour both solutions were cooled to ca. -14 °C using an ice/salt bath. The solution of the anhydride was added all at once to the carbamate solution (exothermic reaction). The ice/salt bath was removed and after 15 min a solution of 8.14 g 1-naphthol (56 mmol) in 40 mL CH_3CN was added to the reaction mixture. The crude reaction was allowed to stir at room temperature for 18 h after which time the pressure was released and the crude mixture was poured into 500 mL 0.5 M aq. HCl. The resulting precipitate was collected by filtration, washed with water and air dried giving 5.68 g (75%) of 1-naphthyl N-methylcarbamate.

Conclusions

A suitable "dehydrating" agent for the highly selective conversion of amines and carbon dioxide into their corresponding isocyanates has been described which eliminates the use of the toxic reagent phosgene. The use cf *o*-sulfobenzoic acid anhydride as the dehydrating agent has also provided a process which recycles the dehydrating agent, thereby potentially eliminating large amounts of salt waste. This technology has also allowed the generation of novel isocyanate materials of interest to the coatings industry, in particular the isocyanate derived from 4-aminomethyl-1,8-diaminooctane (triaminononane), TTI. The use of this particular reagent has also led to a method for the *in situ* trapping of volatile isocyanates (methylisocyanate) which precludes the need for isolation of these toxic reagents.

Literature Cited

1. Waldman, T.E.; McGhee, W.D. *J. Chem. Soc., Chem. Commun.* **1994**, 957-958.
2. McGhee, W.D.; Waldman, T.E. U.S. Patent # 5,189,205, 1993; *Chem Abstr.* **1993**, *119*, 48937p.
3. Riley, D.; McGhee, W.D.; Waldman, T. In *Benign by Design;* Anastas, P.T.; Farris, C.A. Eds.; ACS Symposium Series 577; American Chemical Society, Washington D.C., 1994, pp. 122-132.
4. For examples of some of this work see: McGhee, W.D. U.S. Patent # 5,349,081, 1994; *Chem. Abstr.* **1994**, *121*, 204833a.
5. Clarke, H.T.; Dreger, E.E. In *Org. Syn, Coll Vol I,* Wiley and Sons Inc.; New York, 1941, p 495.
6. Laird, R.M.; Spence, M.J. *J. Chem. Soc. (B)*, **1971**, 1434-1440.
7. Doi, T.; Ide, A.; Kishimoto, Y. U.S. Patent # 4,314,048, 1982; *Chem. Abstr.* **1982**, *96*, 53945a.
8. Higginbottom, H.P.; Hill L.W. Ojunga-Andrew, M. Presented at Third North American Research Conference on Organic Coatings Science and Technology, Hilton Head, SC, Nov. 1994.
9. The relative rates of various carbamates versus acetate in S_N2 type reactions have been discussed in reference 3.
10. Popp, B.; Taunas, N.; Onken, U. U.S. Patent # 3,357,994, 1967; *Chem. Abstr.* **1964**, *60*, 11903d.
11. Nagrasubramanian, K; Chlanda, F.P.; Liu, K.J. *J. Memb. Sci.* **1977**, *2*, 109.
12. Mani, K.M.; Chlanda, F.P.; Byszewski, C.H. Presented at 5th International Symposium on Synthetic Membranes, Tubingen, W. Germany, 1986.
13. For an example of other methods for *in situ* trapping type reactions of methyl isocyanate see: Blaisdell, C.T.; Cordes, W.J.; Heinsohn, G.E.; Kook, J.F.; Kosak, J.R. U.S. Patent # 4,698,438, 1987; *Chem. Abstr.* **1987**, *106*, 101748n.

RECEIVED September 21, 1995

Chapter 5

Clean Oxidation Technologies: New Prospects in the Epoxidation of Olefins

Mario G. Clerici and Patrizia Ingallina

Eniricerche S.p.A., Via Maritano 26, 20097 S. Donato Milanese (Milan), Italy

Two new routes to propylene oxide, based on titanium silicalite catalysis, have been studied. The first route concerns the epoxidation of propylene carried out with *in situ* generation of hydrogen peroxide. The olefin is made to react with air and an alkylanthrahydroquinone in one pot, producing the epoxide in up to 78% yield. The second route is based on the use of the same solvent, *i.e.* a methanol/water mixture, which extracts hydrogen peroxide from the working solution (anthraquinone process) and functions as the epoxidation medium. Organic impurities extracted together with the oxidant do not affect either the rate of reaction or the epoxide purity. Coproduction of chlorinated wastes and organic chemicals, as it occurs in the chlorohydrin and hydroperoxide processes, is avoided. Water is the only coproduct in the new routes. Preliminary evaluations look rather promising and encourage further studies.

In current epoxidation processes, chlorine, hydroperoxides, and peracids are the most commonly used oxidants (*1-2*). Organic and inorganic compounds are coproduced in the reaction, which need to be recycled or disposed off.

In the chlorohydrin route, generally preferred in the epoxidation of C_3-C_6 olefins, stoichiometric amounts of sodium or calcium chlorides are produced by the dehydrohalogenation of intermediate halohydrins. Chlorinated organic by-products, such as halogen ethers and dichlorides, are formed as well in the process, further increasing the quantity of wastes.

In the epoxidation of propylene by the hydroperoxide route, the oxidant is generated by the oxidation of ethylbenzene or isobutane, as are major amounts of corresponding alcohols (1-phenylethanol and *t*-butanol, respectively). The latter, which are also produced in the epoxidation step, are valuable intermedi-

0097–6156/96/0626–0059$12.00/0

ates in the production of styrene and octane enhancers for gasoline, respectively. However, a process designed to produce one single product is generally preferable. *t*-Butyl hydroperoxide is also frequently used in oxidation studies of other olefins. Outstanding results were obtained by Sharpless and coworkers in the asymmetric epoxidation of allyl alcohols (*3*). According to 1991 records, the hydroperoxide processes accounted for 48% of installed propylene oxide capacity and the chlorohydrin process provided 51% of capacity.

Epoxidation by peracids constitutes an indirect use of hydrogen peroxide in commercial processes. C_2-C_3 percarboxylic acids are first generated by the acid catalyzed reaction of H_2O_2 with RCOOH, then used in the epoxidation step. By-product carboxylic acid is then separated and recycled to the peroxidation reactor. Performic acid, which is a hazardous and unstable oxidant generally used in the epoxidation of long chain olefins and vegetable oils, is prepared *in situ* just before use. The cleavage of the oxirane ring, due to acid catalyzed reactions, and the use of chlorinated solvents, preferred for kinetic reasons, are the main drawbacks to the peracid route. Chlorinated compounds are coming under increasingly severe scrutiny due to environmental concerns (*4*). Closely related to the peracids epoxidizing system is the use of acetonitrile/hydrogen peroxide, which coproduces stoichiometric amounts of acetamide (*5*).

Direct use of hydrogen peroxide is a more attractive epoxidation route based on ease and cleanliness of the process. The active oxygen content of H_2O_2, 47 wt%, is much higher than that of organic peracids and hydroperoxides. Water is the only coproduct. In some cases (*vide infra*), hydrogen peroxide can be generated *in situ*, reducing production and transportation costs. Until now, however, the lack of effective catalysts has prevented the direct use of hydrogen peroxide for the epoxidation of olefins, except for a few cases (synthesis of glycidol and epoxysuccinic acid) (*6*). Polar solvents needed to dissolve the reagents, and the presence of water added with the oxidant and formed in the process, inhibit conventional catalysts (*7*). Conversely, performing these reactions at increased temperature or under near anhydrous conditions, greatly decreases the yields of epoxide due to hydrolytic side reactions and H_2O_2 decomposition, or promotes potentially dangerous conditions (*8*).

Titanium Silicalite/Hydrogen Peroxide

New and efficient catalysts for the epoxidation of unactivated olefins with hydrogen peroxide have been discovered in recent years (*9*). These are classified in two groups: phase transfer catalysts (*10*) and metal-substituted zeolites (*11*). Phase transfer catalysts, such as for example $WO_4^{-2}/H_3PO_4/R_4N^+$, are composed by the association of tungstic and phosphoric acids with a quaternary ammonium or phosphonium compound. Titanium silicalite (TS-1) is the most effective catalyst of the second group. The performances of other metal-substituted zeolites in the epoxidation of olefins are still unsatisfactory (*9*).

The epoxidation of propylene with hydrogen peroxide and titanium silicalite occurs under mild conditions (*12*). The reaction can be performed at near room temperature, in methanol or methanol/water solution, in batch-type or continuous reactors (fixed bed or stirred reactor). Yields are generally higher than 90%. Major by-products are propylene glycol, its methyl ether derivatives, and trace amounts of formaldehyde, produced by the solvolysis of the oxirane ring and by the oxidation of the solvent, respectively. Although the rate of epoxidation and the stability of the product are decreased by water, aqueous methanol is normally used for practical reasons, since water is both added with the oxidant and formed during the reaction. Below 50wt% content its effects on kinetics and epoxide yields are rather moderate (*12*). Yields become almost quantitative (*ca* 97%) when a basic salt, such as sodium acetate, either is used to pretreat TS-1 or is directly added into the epoxidation medium (*12*). The amount depends on operating conditions and is generally kept low to avoid the formation of anionic TS-1 peroxides, which are inert in the epoxidation of olefins (*14*).

A dual role is played by methanol, which is both a solvent for reagents and products and a cocatalyst. Studies on the epoxidation mechanism, on the formation of TS-1 peroxides and on acid properties of TS-1/H_2O_2 system (*12-15*), suggest that methanol takes part in the reaction mechanism by promoting the formation of the active species (Figure 1). Accordingly, reaction kinetics reaches maximum efficiency in the presence of methanol (*12, 13*). Initial turnover frequencies of 1-2 s^{-1} have been observed at 40°C, in 92wt% methanol. Other polar solvents, even pure water, are usable provided that a decrease in the rate of reaction and in the yields is tolerated (*12*).

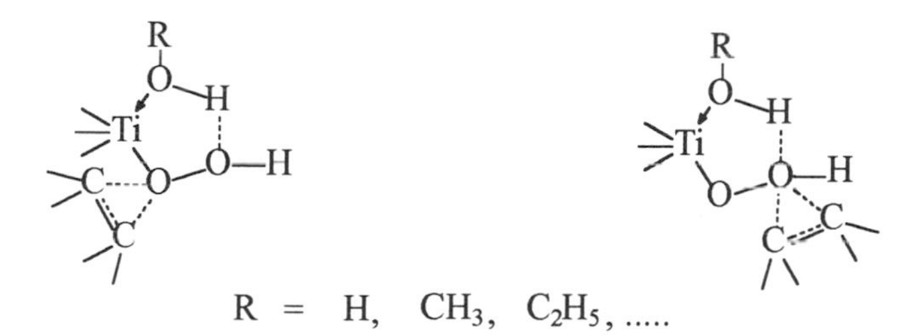

Figure 1. Proposed active species and epoxidation mechanism (*13*).

The use of technical grade solutions of hydrogen peroxide does not affect the yields or the rate of the reaction, because molecular sieve properties of the catalyst prevent bulky molecules, which might be present as impurities in the H_2O_2 solution, from diffusing inside TS-1 channel system where active sites are located. Thus, hydrogen peroxide can be extracted from alkylanthraquinone working solution and used without further purification in the epoxidation of propylene (Clerici, M. G.; D'Alfonso, A., unpublished results). It does not need to be concentrated since, even at 1wt% H_2O_2 (or lower), both the rate of epoxidation and the yields are high (>90%). These considerations are most important for oxidations based on hydrogen peroxide produced *in situ*.

Longer chain olefins are similarly epoxidized, with yields in the range of 80-98% (*13*). The rate of reaction strongly depends on structural features of the olefin including chain length, presence of substituents, and position and steric configuration of the double bond (*13*). As a result, a different order of reactivity is shown by TS-1 as compared to other epoxidation catalysts: α-olefin > internal olefin, linear olefin > branched and cycloolefin, linear-C_n > linear-C_{n+1}. *Cis*-2-butene reacts 16 times faster than the *trans*-isomer (*13*).

Despite the presence of an electron withdrawing substituent, the epoxidation of either allyl chloride or allyl alcohol using TS-1 is fast and selective at 40°C (*13*). The new route appears to be advantageous over existing technologies based on the halohydrin process and on WO_4^{-2}/H_2O_2, respectively.

Titanium Silicalite/Molecular Oxygen (*in situ* H_2O_2)

Although the production propylene oxide *via* H_2O_2 would be very attractive because of the cleanliness and the ease of reaction, the method has drawbacks. Hydrogen peroxide is still a relatively expensive reagent. When producing low molecular weight compounds, a relatively high amount of oxidant is needed. For example, on a stoichiometric basis 1 kg of H_2O_2 is required to produce 1.7 kg of propylene oxide. The shipment of large amounts of concentrated hydrogen peroxide may also be an issue.

While economically viable solutions can be found for these problems, such as on site production of hydrogen peroxide and employing the latter as technical grade dilute solution, another option considered involved using molecular oxygen of air directly. More specifically, the high activity of TS-1 in very dilute solutions of hydrogen peroxide was taken advantage of to study the oxidation of saturated and unsaturated hydrocarbons based on *in situ* generation of H_2O_2 with H_2 and O_2 (*16-18*). To this purpose, Pd or Pt was supported on TS-1 or an organic redox system was added to the reaction medium.

The metal modified TS-1 is suited mostly to carry out the oxidation of paraffins since the presence of acids, required to favour hydrogen peroxide formation (*19*), is incompatible with the stability of the oxirane ring (*15*). The selection of the organic redox system, among those known in the literature to generate hydrogen peroxide by the oxidation with molecular oxygen (*6*), is made on the basis of molecular size to avoid any catalyzed degradation. The molecular sieve structure of TS-1 prevents all compounds having cross section larger than *ca* 0.55 nm from diffusing inside TS-1 channels and, therefore, from reacting or interfering with reactions occurring at Ti-sites. Alkylanthraquinones, which are currently used in commercial processes for hydrogen peroxide production, are good candidates for the epoxidation of propylene with *in situ* generated H_2O_2 (equations 1-2). Other organic compounds (QH_2) can be similarly used (*17*).

$$QH_2 + O_2 + \text{propylene} \xrightarrow{\text{TS-1}} Q + \text{propylene oxide} + H_2O \quad (1)$$

$$Q + H_2 \longrightarrow QH_2 \quad (2)$$

The carrier, QH_2, can be shown to react in one pot with molecular oxygen and propylene to produce propylene oxide, water, and the corresponding oxidized precursor Q. The latter is separately hydrogenated to close the cycle of reactions (equations 1-2). Actually, equation 1, hydrogen peroxide is produced in a first step in the reaction medium by QH_2 and O_2 and, in a second step, reacts with propylene at Ti-sites as previously reported (equations 3-4) (*12-13*).

$$QH_2 + O_2 \longrightarrow H_2O_2 + Q \quad (3)$$

$$H_2O_2 + \text{propylene} \xrightarrow{\text{TS-1}} \text{propylene oxide} + H_2O \quad (4)$$

Molecular dimensions restrict QH_2 and Q from diffusing inside TS-1 channels and from possible degradation. A specific redox system (QH_2/Q), constituted by an alkylanthrahydroquinone/alkylanthraquinone couple, is illustrated in equation 5.

$$\text{alkylanthrahydroquinone (OH, OH, R)} + O_2 + \text{propylene} \xrightarrow{\text{TS-1}} \text{alkylanthraquinone (O, O, R)} + \text{propylene oxide} + H_2O \quad (5)$$

The solvent plays a major role in the process. Its selection is strictly dependent on the properties of both the catalyst and the redox system. More specifically, it must be stable to oxidative degradation by TS-1, which means that it should be chemically inert or sterically hindered. The solubility of all the components of reaction mixture has to be high, to minimize the volume of the working solution. This is a difficult task to accomplish, since solubility properties of reagents and products differ significantly. Third, the solvent should promote epoxidation kinetics (*12*).

Solubility and reactivity tests, performed on potential solvents, showed that one single solvent could not fulfil all the above requirements (Table I). Methylnaphtalene and diisobutylcarbynol were good solvents for alkylanthraquinones (Q) and for alkylanthrahydroquinones (QH_2), respectively. Diisobutyl ketone dissolved both compounds, however, at relatively low concentrations. Methanol has a positive effect on kinetics since it promotes the formation of active species (Figure 1), but it shows poor solvent properties for the redox system.

Table I. Epoxidation of Propylene with *in situ* H_2O_2. Solvent Properties [a].

Solvent	*Q*	*QH_2*	*Kinetic*
Methylnaphtalene	+ +	-	- -
Diisobutyl ketone	+	+	- -
Diisobutylcarbynol	-	+ +	- -
Methanol	- -	+	+ +

[a] The symbols + and - are used as qualitative indexes of each solvent properties.

Although a number of alkyl-substituted anthraquinones are potential candidates in the *in situ* generation of H_2O_2, ethyl- and *t*-butyl-anthraquinone have been used in preliminary experiments (equation 5). Summaries of these reactions are provided in Table II and include solvent composition, quinone concentration, and yields. The yields of propylene oxide, based on starting alkylanthrahydroquinones, were 78% and 62%, respectively, when ethylanthrahydroquinone or a mixture of *t*-butyl- and ethyl-substituted anthrahydroquinone were reacted, at room temperature, with propylene and oxygen (from air) in the presence of TS-1. The higher solubility of the second redox system significantly improves the concentration of the product.

Table II. Epoxidation of Propylene with *in situ* H_2O_2[a].

Solvent Composition			*Quinone*		*PO*
MEN (vol%)	*DIBC (vol%)*	*CH_3OH (vol%)*	*R* [b]	*Conc. (M)*	*Yield (%)*
22	68	10	C_2H_5	0.13	78
40	50	10	C_2H_5(45%)+*t*-C_4H_9(55%)	0.22	62

[a] PO, propylene oxide; MEN, 1-methylnaphtalene; DIBC, diisobutylcarbynol.
[b] See equation 5.

The yields shown by Table II are somewhat lower than those already reported for the propylene oxide synthesis using preformed hydrogen peroxide (*12*). However, due to the preliminary nature of the experiments carried out and the complexity of the reactions involved, the results are susceptible of further improvement.

Scheme I shows a simplified block diagram illustrating the four main steps of a new route to propylene oxide production. In the first step, an alkylanthrahydroquinone, propylene and air react through a series of reactors producing propylene oxide, a minor amount of solvolysis products and water. Propylene oxide is separated by distillation and recovered. In the next step, methanol, propylene glycol and its methyl ether derivatives are extracted with water and purified. The remaining organic phase passes to the alkylanthraquinone purification/hydrogenation step and finally is fed with methanol, back to the epoxidation reactors. The regeneration and purification of the working solution are not shown in Scheme I.

Scheme I

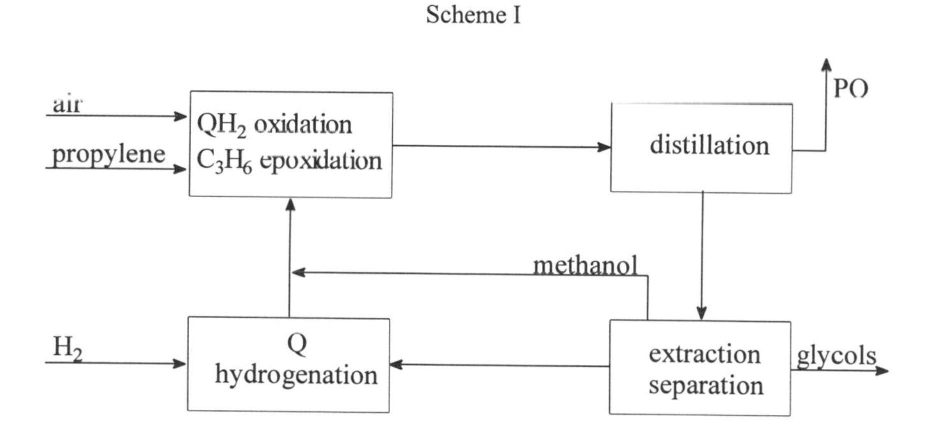

A preliminary economic estimate has been carried out, on the basis of the results of Table II and Scheme I, and compared to the chlorohydrin and hydroperoxide routes. Net production costs of the three processes are fairly close to each other, with the *in situ* hydrogen peroxide route being the most advantageous. Although the economic estimate looks rather promising, the *in situ* H_2O_2 route is based on results still at the laboratory scale prior to optimization, whereas the other two are already commercial processes. Nonetheless, the comparison of the three methods indicates that the production of propylene oxide by *in situ* hydrogen peroxide and TS-1 catalysis might eventually be a viable process and, therefore, is worthy of further study.

The molecular sieve properties of TS-1 suggest an alternative route, shown by Scheme II. It is based on the use of the same solvent mixture, *i.e.* methanol/water, both to extract H_2O_2 from the working solution and to carry out the epoxidation of the olefin (*18*). Thus, the cost of purifying hydrogen peroxide is eliminated and extra water is not required to be added to the epoxidation reactors with the oxidant, since hydrogen peroxide is extracted directly by the epoxidation medium. Only water produced by the consumption of hydrogen peroxide needs to be eventually removed from the reaction stream.

Scheme II

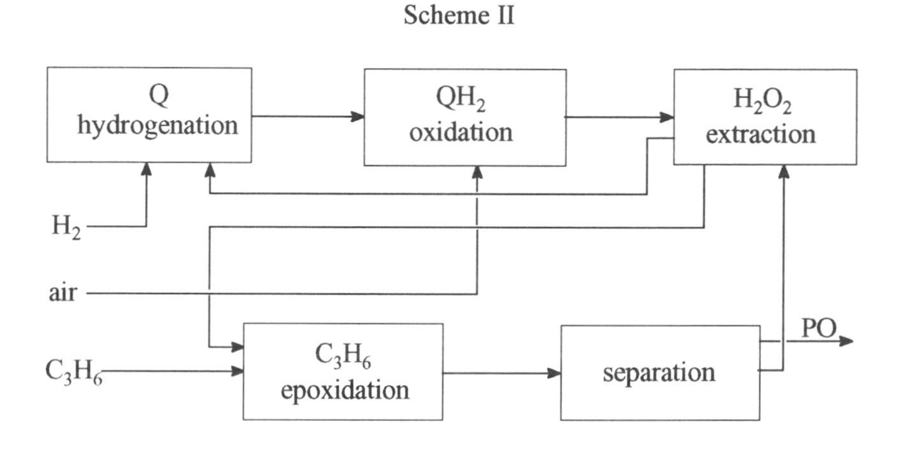

The feasibility of the process shown by Scheme II relies on the efficiency of hydrogen peroxide extraction, while keeping all the components of the working solution, particularly the alkylanthraquinone, from migrating into the epoxidation solvent and *viceversa.* This is accomplished by choosing for the extraction/epoxidation solvent a methanol/water ratio suitable to the composition of the working solution used. As an example, Table III shows the results of the extraction of hydrogen peroxide, using a 52wt% methanol/water mixture, from a solution obtained by the oxidation of 2-ethylanthrahydroquinone dissolved in diisobutyl ketone. Both the amount of methanol in the working solution and the components of the latter passing into the extracting solvent are negligible.

Table III. Extraction of H_2O_2. Component distribution in diisobutyl ketone (DIBK) and CH_3OH/H_2O phases [a].

DIBK Phase		*CH_3OH/H_2O Phase*		
CH_3OH	Q	H_2O_2 [b]	Q	DIBK
16 g/l	48 g/l	40 g/l	0.04 g/l	6.6 g/l

[a] 2-ethylanthrahydroquinone (QH_2) in 25 ml of diisobutyl ketone is oxidized with air and then extracted with 2x2ml CH_3OH/H_2O solution (CH_3OH 52wt%), at 25°C.

[b] Combined extracts.

The solution was used directly in the epoxidation step without any purification, verifying that the small amounts of organic materials extracted together with hydrogen peroxide do not interfere with the subsequent epoxidation process (*18*). The result is in good agreement with those of a previous study in which the usability of technical grade solutions of hydrogen peroxide, coupled with the operation of a continuous stirred reactor, was demonstrated. Both the rate of the reaction and the purity of the product were not affected by the presence of organic impurities originating from the working solution of a commercial plant (Clerici, M. G.; D'Alfonso, A.; Maspero, F., unpublished results).

A preliminary evaluation shows that a process based on Scheme II is advantageous with respect to the method using purified preformed hydrogen peroxide (*12*). It requires, however, that both the hydrogen peroxide and the propylene oxide plants are located in close proximity to each other.

Conclusions

Titanium silicalite is an effective catalyst for the synthesis of propylene oxide with air and hydrogen in the presence of carriers able to generate *in situ* hydrogen peroxide. While alkylanthraquinones were demonstrated to be effective carriers in this study, other organic carriers are also eligible for this purpose. The reactions involved occur at near room temperature, are selective, and employ readily available feedstocks. The coproduction of major amounts of other chemicals is avoided. As an alternative, hydrogen peroxide produced by the anthraquinone process can be extracted by means of the epoxidation solvent, a methanol-water mixture, and used without further purification in the epoxidation step. Although both routes are based on results still at the laboratory stage, preliminary economic estimates show that these might constitute viable alternatives to present technologies.

Both Schemes I and II are applicable to other reactions catalyzed by TS-1. C_4-C_5 olefins can be epoxidized according to Scheme I. Scheme II is adaptable to carry out the epoxidation of a range of olefins, the ammoximation of cyclohexanone, the hydroxylation of aromatics and the oxyfunctionalization of paraffins (*18*).

An issue which deserves further mention is the environmentally friendly nature of TS-1/H_2O_2 system. It involves the use of a safe silica based catalyst, titanium silicalite, and a reagent, hydrogen peroxide, which yields water as the coproduct. This holds for the *in situ* route illustrated in Scheme I and also for the epoxidation of propylene with preformed hydrogen peroxide, either used as an aqueous solution (*12*) or extracted by means of the epoxidation solvent (Scheme II). Hazardous chemicals, such as chlorine, performic or other organic peracids, are not required in the process. The disposal of chlorinated salts or the recycle of brine (chloroydrin process) and any possible burden resulting from the coproduction of other chemicals (styrene and *t*-butanol in the hydroperoxide route) are eliminated. The liquid phase oxidation of isobutane and ethylbenzene with air under pressure and at high temperature, to produce corresponding hydroperoxides needed in the epoxidation step, is compared with the mildness of the reactions involved in the TS-1/H_2O_2 system. Minor safety risks associated with the shipment and storage of concentrated hydrogen peroxide solutions are avoided since H_2O_2 is produced directly *in situ* or is extracted and employed as a dilute solution.

Literature Cited

1. Sienel, G.; Rieth, R.; Rowbottom, K. T. In *Ullmann's Encyclopedia of Industrial Chemistry,* 5th ed; VCH Publishers: New York, NY, 1987; Vol. A9, p. 531.
2. Kahlich, D.; Wiechern, U.; Lindner, J. In *Ullmann's Encyclopedia of Industrial Chemistry,* 5th ed; VCH Publishers: New York, NY, 1987; Vol. A22, p. 239.
3. Hanson, R. M.; Sharpless, K. B. *J. Org. Chem.,* **1986**, *51,* 1922.
4. Hileman, B.; Hanson, D. *Chem. Eng. News,* **1994**, *7 Feb.*, 4.
5. Payne, G. B.; Deming, P. H.; Williams, P. H. *J. Org. Chem.,* **1961**, *26,* 659.
6. Goor, G. In *Catalytic Oxidations with Hydrogen Peroxide as Oxidant;* Strukul, G., Ed.; Kluwer Academic Publishers: Dordrecht, 1992; p.13.
7. Sheldon, R. A. *J. Mol. Catal.,* **1983**, *7,* 107.
8. Schirmann, J. P; Delavarenne, S. Y. *Hydrogen Peroxide in Organic Chemistry;* Edition ed Documentation Industrielle: Paris, 1979.
9. Clerici, M. G., In *Heterogeneous Catalysis and Fine Chemicals III*; Guisnet, M., Barbier, J., Barrault, J., Bouchoule, C., Duprez, D., Pérot, G., Montassier, C., Eds; Studies in Surface Science Catalysis; Elsevier: Amsterdam, 1993, Vol.78; p. 21.
10. Venturello, C.; Alneri, E.; Ricci, M. *J. Org. Chem.,* **1983**, *48,* 3831.
11. Taramasso, M.; Perego, G.; Notari, B. U. S. Patent 4410501, 1983.
12. Clerici, M. G.; Bellussi, G.; Romano, U. *J. Catal.,* **1991**, *129,* 159.
13. Clerici, M. G.; Ingallina, P. *J. Catal.,* **1993**, *140,* 71.
14. Clerici, M. G.; Ingallina, P. In *Proceedings from the Ninth International Zeolite Conference;* von Ballmoos, R., Higgins, J. B., Treacy, M. M. J., Eds.; Butterworth-Heinemann: Boston, MA, 1993, Vol. 1; p. 445.
15. Bellussi, G.; Carati, A.; Clerici, M. G.; Maddinelli, G.; Millini, R. *J. Catal.,* **1992**, *133,* 220.
16. Clerici, M. G.; Bellussi, G. U. S. Patent 5235111, 1993.
17. Clerici, M. G.; Ingallina, P. U. S. Patent 5221795, 1993.
18. Clerici, M. G.; Ingallina, P. U. S. Patent 5252758, 1993.
19. Gosser, L. W. U. S. Patent 4681751, 1987.

RECEIVED November 7, 1995

Alternative Syntheses and Reagents

Chapter 6

Dimethylcarbonate and Its Production Technology

F. Rivetti, U. Romano[1], and D. Delledonne

Research and Development Department, EniChem Synthesis, Via Maritano 26, 20097 S. Donato, Milanese (Milan), Italy

Dimethylcarbonate (DMC), a non toxic, non polluting chemical, is achieving increasing importance as a versatile intermediate and product. Of the several synthetic routes to DMC that are available, oxidative carbonylation of methanol appears to be the most attractive. This route, coupled with the use of cuprous chloride as the catalyst led to the development of an industrially viable process, successfully used by EniChem Synthesis. Recent developments allow easy scale up to large capacity plants.

Dimethylcarbonate (DMC) is a versatile and environmentally innocuous material for the chemical industry*(1)*, as indicated by its toxicological and ecotoxicological properties, reported in Tables I and II.

Table I Toxicological Properties of DMC

Acute toxicity (ORAL, RAT)	LD_{50} 13.8 g/kg
(SKIN, RAT)	LD_{50} > 2.5 g/kg
(INHAL., 4h, RAT)	LD_{50} 140 mg/l
MUTAGENIC PROPERTIES (Ames, DNA REPAIR, OECD 473, OECD 476)	NEGATIVE
IRRITATING PROPERTIES (RABBIT, EYE, SKIN)	NEGATIVE
90 dd TOXICITY (ORAL, RAT, OECD 408)	NEL 500 mg/kg/d

NEL = No-effect level

[3]Current address: EniChem, I.G. Donegani, Via Fauser 4, 28100 Novara, Italy

0097–6156/96/0626–0070$12.00/0

Table II Ecotoxicological Properties of DMC

Biodegradation (OECD 301 C)	> 90% (28 dd)
Acute toxicity on fish (Leuciscus idus, OECD 203)	NOEC 1000 mg/l
Acute toxicity on aerobic waste water bacteria (OECD 209)	EC_{50} > 100 mg/l

NOEC = No-effect concentration

DMC has been proposed as a safe alternative to the use of toxic, chlorinated intermediates, such as phosgene in carbonylation reactions and methyl chloride and dimethyl sulfate in methylation reactions.

Phosgene, dimethyl sulfate and methyl chloride are typical highly reactive chemicals used since the beginning of the chemical industry. The cost of their utilization, however, is dramatically increasing owing to the growing safety measures during their production, transportation, storage and use. Phosgene has been used as a war gas because of its toxicity. Dimethyl sulfate is a suspect human carcinogen, besides being extremely toxic. Their use in carbonylation and methylation reactions brings about the formation of stoichiometric amounts of by-products such as inorganic chlorides or sulfates, generally in form of organically contaminated aqueous streams. The problems raised by their use are noticeable because of the large consumption of these chemicals. Phosgene yearly consumption ranks as high as 2 million tons, mainly for production of isocyanates for polyurethanes, polycarbonates and agrochemicals. Some ten thousand tons of methylating agents are used in the production of phenolic ethers, quaternary ammonium salts and fine chemicals.

DMC has been proposed as a non-toxic, non-polluting solvent *(2)*. It is being considered as a component of reformulated fuels, owing to its high oxygen content and good blending properties *(3-4)*.

The EniChem Project

A project aimed at the development of the industrial production of DMC and its derivatives was initiated in the 1980s by EniChem Synthesis (ECS), a branch of EniChem Group. This project is an excellent example of the current effort by the chemical industry toward product and process innovation aimed at avoiding environmental concerns. The goals of this effort are achieved through the following guidelines: substituting toxic, dangerous, highly reactive chemicals with less reactive, less harmful, more selective building blocks; activating selective chemical reactions by proper catalysis; substituting old technologies with new ones characterized by the reduction of by-products and/or their easy separation and recycling.

These guidelines are an important element of product & process development by the future chemical industry.

With an interest in industrial use of DMC in large volume applications, a major task of the ECS was to develop a process suitable for large scale production of DMC that would start from easily available and inexpensive raw materials.

In addition to the DMC synthesis, ECS also has developed several new chemical processes and products that are based on DMC. These processes show a series of favorable features including the use of a harmless reagent, absence of solvents, absence of waste water, easily disposable (CO_2) or recyclable (CH_3OH) by-products, as depicted in Table III.

Table III Comparison Between DMC and Phosgene or Dimethyl Sulfate (DMS) Based Processes

Phosgene or DMS	**DMC**
Dangerous reagents	Harmless reagent
Use of solvent	No solvent
Waste water treatment	No waste water
NaOH consumption	————
By- products: NaCl, Na_2SO_4	By- products: CH_3OH, CO_2
Exothermic	Slightly or not exothermic

DMC Production

The traditional method for synthesizing DMC is the reaction of CH_3 OH with phosgene (eq.1).

$$COCl_2 + 2CH_3OH \longrightarrow CH_3OCOOCH_3 + 2HCl \qquad (1)$$

To avoid the use of the extremely toxic chemical, phosgene, several alternative routes have been considered based on the use of carbon monoxide (oxidative carbonylation) or carbon dioxide (through direct or indirect carboxylation). The two most attractive routes for industrial syntheses were found to be oxidative carbonylation of methanol in the presence of a suitable catalyst (eq.2)

$$2CH_3OH + CO + 1/2\, O_2 \longrightarrow CH_3OCOOCH_3 + H_2O \qquad (2)$$

and carboxylation/transesterification of ethylene oxide to DMC via ethylene carbonate *(5)* (eq. 3a, 3b).

$$\text{H}_2\text{C}\overset{\text{O}}{-}\text{CH}_2 + \text{CO}_2 \longrightarrow \text{(ethylene carbonate: cyclic } \text{OC(=O)O–H}_2\text{C–CH}_2\text{)} \quad (3a)$$

$$\text{(ethylene carbonate)} + 2\text{CH}_3\text{OH} \longrightarrow \text{CH}_3\text{OCOOCH}_3 + \text{HOCH}_2\text{CH}_2\text{OH} \quad (3b)$$

A major drawback of the carboxylation/transesterification of ethylene oxide is the co-production of ethylene glycol.

On the basis of the technical and economical considerations, the methanol oxidative carbonylation process (eq. 2) was selected.

Methanol Oxidative Carbonylation

Oxidative carbonylation is known to take place in the presence of a number of metal ions and complexes. Most of the metal ions suitable in oxidizing CO in water to CO_2 *(6)* are also active in the oxidative carbonylation of alcohols because of the similarity of these two basic reaction pathways (eq. 4a, 4b and 5a, 5b).

$$M^{n+} \xrightarrow[-H^+]{CO,\ H_2O} [M\text{-}COOH]^{(n-1)+} \longrightarrow M^{(n-2)+} + CO_2 + H^+ \quad (4a)$$

$$[M\text{-}COOH]^{(n-1)+} \xrightarrow{M^{n+}} 2M^{(n-1)+} + CO_2 + H^+ \quad (4b)$$

$$M^{n+} \xrightarrow[-H^+]{CO,\ ROH} [M\text{-}COOR]^{(n-1)+} \xrightarrow{ROH} M^{(n-2)+} + (RO)_2CO + H^+ \quad (5a)$$

$$[M\text{-}COOR]^{(n-1)+} \xrightarrow[ROH]{M^{n+}} 2M^{(n-1)+} + (RO)_2CO + H^+ \quad (5b)$$

An oxidant is necessary to re-oxidize the metal ion, once in its low valence state, back to its original oxidation state and establish a catalytic cycle. The reoxidation is preferably carried out in-situ. Interestingly, the oxidative carbonylation also takes place in the presence of non-metal redox pairs such as Se^{2-}/Se^0 *(7)* and Br^-/Br_2 *(8)*.

Important technical parameters in choosing a suitable catalytic system are the easiness of low valence metal reoxidation by inexpensive oxidants (such as molecular oxygen) and the selectivity toward competitive reactions, mainly alcohol oxidation and formation of CO_2. The latter is co-produced, under catalytic conditions, in the presence of water, generated when oxygen is used for the simultaneous catalyst reoxidation; see equation 2.

Particular attention has been devoted to the catalytic activity of palladium (Pd^0/Pd^{2+}) salts, mostly because of their high reactivity in mild conditions. The reaction can be driven toward the formation of oxalate or carbonate *(9-11)*. The reoxidation of Pd^0 salts with oxygen can be easily performed in-situ if a suitable co-catalyst is used. Copper has been the most widely proposed. Unfortunately, the formation of water during the reoxidation process heavily interferes with the catalytic cycle. Therefore, non-conventional oxidants like peroxides *(12)* or processes involving a separate reoxidation step have been proposed. A vapor phase two-step process based on palladium catalyzed carbonylation of methyl nitrite has been developed on an industrial scale in Japan *(13)*.

Methyl nitrite is generated in a separate step by reaction of CH_3OH, NO and O_2 (eq. 6a, 6b).

$$2CH_3ONO + CO \longrightarrow CH_3OCOOCH_3 + 2NO \quad (6a)$$

$$2NO + 2CH_3OH + 1/2O_2 \longrightarrow 2CH_3ONO + H_2O \quad (6b)$$

Copper Catalysis

Amine - copper(II) complexes are very efficient in the production of DMC in the methanol/CO system*(14)* and (F. Rivetti, D. Delledonne, unpublished results) (Table IV). However, their practical use is hampered because this catalytic system fails to produce DMC efficiently under technical conditions. A major drawback of this system is the sensitivity to the presence of water, co-produced when using oxygen for in-situ reoxidation of the catalyst, as outlined before. Water has a deep negative influence on the catalyst activity and reaction selectivity, resulting in slow reaction rate and high CO_2 production. Moreover, under oxidative condition, the amino ligands undergo degradation resulting in a poor catalyst life.

Amine-copper complexes are able to produce DMC even at room temperature under carbon monoxide at atmospheric pressure. Simple copper salts, like CuCl, require more drastic reaction conditions to effect the formation of DMC *(15)*, e.g. temperature starting from about 70 °C. On the other hand they do not suffer the problems associated with the use of copper complexes, so that, a catalytic cycle with simultaneous reoxidation by O_2 is easily established using CuCl.

According to a simplified scheme, the reaction proceeds initially through CuCl oxidation by O_2 to cupric methoxychloride (eq. 7), a polymeric species insoluble in the reaction medium *(16)*.

$$2CuCl + 2CH_3OH + 1/2O_2 \longrightarrow 2Cu(OCH_3)Cl + H_2O \quad (7)$$

The latter is reduced by carbon monoxide to DMC, thus restoring CuCl and completing the catalytic cycle (eq. 8).

$$2Cu(OCH_3)Cl + CO \longrightarrow 2CuCl + CH_3OCOOCH_3 \quad (8)$$

Table IV DMC Formation by Copper Amine Complexes $L_nCu(OCH_3)$ Cl Reduction

L (mol/l)	Time (h)	Yield (%)
PY (3.3)	2.5	100
DMPY (3.3)	5	tr
TMEDA (1.1)	2.25	88
TEEDA (1.1)	5	16
TMMDA (1.1)	3.5	29
TMPDA (1.1)	3.5	12
BIPY (1.1)	0.16	90
TEA (2.2)	5	0

$Cu(OCH_3)Cl$ 1.10 mol/l, pCO 0.8 MPa, T 25°C
py= pyridine, DMPY=2,6 dimethylpyridine, TMEDA=tetramethyl - ethylenediamine, TEEDA = tetraethylethylenediamine, TMMDA = tetramethyldiaminomethane, TMPDA=tetramethyl-1,3-diaminopropane, BIPY = 2,2' bipyridyl, TEA = triethylamine.

This simplified scheme is not adequate since it was observed that the reduction of $Cu(OCH_3)Cl$ by CO is promoted by addition of even small amounts of a soluble cupric species, such as $CuCl_2$, or water, which results in the availability of Cu(II) in solution, by hydrolysis *(17-18)* (Table V). Therefore, cupric methoxychloride cannot be considered the true reaction intermediate.

In support of this observation, a methanol dispersion of cupric methoxychloride synthesized by CuCl oxidation with oxygen in methanol presents a ESR spectrum with a resolved signal typical of cupric ions in solution superposed to the enlarged signal characteristic of the polymeric species. Such data have been attributed to the hydrolysis reaction between $Cu(OCH_3)Cl$ and water, a co-product of the reaction*(15)*.

Moreover, in systems containing higher water concentrations, (5-10 % by weight) such as those resulting in technical conditions from direct O_2 reoxidation when the reaction is performed at significant methanol conversion, exhaustive hydrolysis of copper methoxychloride leads to a number of soluble and insoluble cupric species containing chloride and/or hydroxide counterions in different ratios, particularly $CuCl_2$ and $Cu_4Cl_2(OH)_6$ (atacamite). This reaction system results in DMC and cuprous chloride formation by reduction under CO.

Table V DMC Formation by $Cu(OCH_3)Cl$ Reduction

H_2O, mol/l	Yield %
0.3	22
0.6	42
1.1	85
$CuCl_2$ mol/l	Yield %
0.0084	29
0.042	55
0.084	77

$Cu(OCH_3)Cl$ 1.68 mol/l, pCO 1.2 MPa, T 75 °C, Time 4h

In order to gain further insight into the reaction pathway, $Cu(OCH_3)Cl$ was prepared by CuCl oxidation with oxygen in methanol, isolated, then reacted under carbon monoxide in ethanol or ethanol/ alkanol mixtures. The results are reported in Table VI.

Table VI $Cu(OCH_3)Cl$ Reduction by CO in Alcohols

liquid medium	**Yield to dialkyl carbonate**[b] **$\%_w$**	**Selectivity to carbonates (%)**
Ethanol	22	Diethylcarbonate (100)
Ethanol/methanol[a]	52	DMC (51) Ethylmethylcarbonate (40) Diethylcarbonate (9)
Ethanol/ 1-propanol[a]	14	Diethylcarbonate (31.5) Ethyl,1-propyl carbonate (44) Bis(1-propyl) carbonate (24)
Ethanol/ 2-propanol[a]	6	Diethylcarbonate (83) Iso-propyl ethyl carbonate (17)

[a] 1/1 molar mixture
[b] calculated on $Cu(OCH_3)Cl$
Experimental conditions: $Cu(OCH_3)Cl$ 5 mmol, alcohol 200 or 100 + 100 mmol, Temperature 80 °C, CO pressure 3.5 MPa, rection time 5h.

Under these experimental conditions, transesterification (exchange of a methoxy for an alkoxy group) in $Cu(OCH_3)Cl$ was proven not to occur to a significant degree.

In addition, it was observed that methanol formation closely accompanied the production of the carbonate. Finally, dialkyl carbonates yields and selectivities indicate a decrease of alkanol reactivity in the series: methanol >> ethanol >1-propanol >> 2-propanol.

Although the reaction pathway is not known in detail, these results support an attack of an alcohol molecule on carbon monoxide coordinated to a cupric species in solution bearing chloride (and possibly hydroxo) ligands, for the generation of the alkoxy carbonyl intermediate (see equation 5). The hydroxo ligands buffer the system acidity.

ECS technology for DMC production

The industrial technology for DMC production developed by ECS is based on the one-step, liquid phase, oxidative carbonylation of methanol in the presence of cuprous chloride using oxygen as a direct oxidant *(19)*. This technology is used by ECS for the production of DMC at its Ravenna facilities. The main features of the process have already been outlined *(1,15)*. Technically the reaction is performed by continuously feeding the reactor with liquid methanol and a gaseous stream containing carbon monoxide and oxygen. Reaction conditions are in the range of 120-130 °C and 2-3 MPa. Water concentration can be as high as 10-12% by weight in the reaction medium. The fed oxygen is completely consumed by the reaction avoiding problems deriving from the formation of explosive mixtures in the reactor.

The process starts from widely available and inexpensive raw materials. Mixtures of carbon monoxide and hydrogen can be conveniently employed, since H_2 does not have any effect on the reaction *(20)*. The process is an example of a clean technology. For example, high selectivity on methanol is achieved, water is the only co-product of the reaction, and CO_2 (formed from CO as the only by-product in substantial amount) can be efficiently re-utilized as a carbon source in the CO generation process. The high selectivity translates into high product purity; DMC produced by this process is characterized by the practical absence of chlorinated impurities.

Recent developments

Alternative technological implementations of the oxidative carbonylation of methanol include the development of vapor phase processes over heterogeneous supported catalysts *(21-22)*, such as $CuCl_2$ on active carbon.

A major advantage of vapor phase processes is the easy separation of reaction products from the catalytic system. Selectivity and catalyst deactivation and regeneration procedures, however, appear to be more critical compared to the liquid phase process.

Recently ECS introduced the concept of reactor-evaporator in the liquid phase, CuCl based, DMC technology*(23)*, affording a process that maintains the advantages of the liquid phase process with respect to selectivity and catalyst stability and greatly simplifies the separation of the product from the catalytic system in a manner similar to vapor-phase processes. In this technology, the catalyst is kept inside the reactor and the vaporized products are removed from the reaction system with the gases leaving the reactor. This allows the use of higher catalyst concentration and therefore higher productivities in DMC production.

Cobalt catalyzed DMC production

The possibility of using, instead of CuCl, a non-corrosive catalytic system appears to be very attractive, since the use of a catalyst containing copper and chloride results in severe requirements when selecting plant construction materials because of its corrosivity.

Oxidative carbonylation of amines to carbamates and ureas in the presence of cobalt complexes was recently reported *(24-25)*. We have found that Co(II) complexes bearing nitrogen and oxygen-donor ligands such as carboxylates, acetylacetonates and Schiff bases produce DMC with high selectivity in methanol under CO and O_2. (Table VII) *(26-27)*.

Table VII DMC Formation by Cobalt Complexes

Catalyst mol/l	Time h	Temp. °C	PCO MPa	PO_2 MPa	DMC g/l	Turnover
$Co(CH_3COO)_2 \cdot 4H_2O$ (0.16)	2.5	100	2	1	35	2.4
$Co(acac)_2$ (bipy) (0.08)	5	100	2	1.0	71	10
$Co(pic)_2 \cdot 4H_2O$ (0.065)	5	120	1.6	0.4	75	13
$[Co(Salen)]_2 \cdot H_2O$ (0.029)	7	120	1.6	0.4	301	58

acac = acetylacetonate, salen = 1,6 bis (2-hydroxyphenyl)-2,5-diaza-1,5-hexadiene
pic = 2-pyridinecarboxylate

In the case of $[Co(Salen)]_2 H_2O$ (Salen = 1,6 bis (2-hydroxyphenyl) -2,5-diaza -1,5- hexadiene), a methoxycarbonyl derivative can be formed and isolated from the methanol, CO, and O_2 reaction system at room temperature (eq. 9) *(28)*.

$$[Co\ (Salen)]_2 \cdot H_2O \xrightarrow{CH_3OH,\ CO,\ O_2} 2[Co(Salen)COOCH_3] \quad (9)$$

We have observed that [Co(Salen)$COOCH_3$] can be isolated in quantitative yield from oxidative carbonylation experiments using [Co(Salen)]$_2$· H_2O as the catalyst. Furthermore, [Co(Salen)$COOCH_3$] behaves as [Co(Salen)]$_2$·H_2O in promoting the reaction when it is used as starting material. By heating [Co(Salen)$COOCH_3$] in CH_3OH or EtOH at 100°C under nitrogen, DMC or ethylmethyl carbonate is produced in about 50% yield from the catalytic system (eq. 10a, 10b).

$$[Co(Salen)COOCH_3] \xrightarrow{100\,°C,\ N_2,\ MeOH} Co(Salen)\cdot CH_3OH + CH_3OCOOCH_3 \quad (10a)$$

$$[Co(Salen)COOCH_3] \xrightarrow{100\,°C,\ N_2,\ EtOH} Co(Salen)\cdot EtOH + CH_3CH_2OCOOCH_3 \quad (10b)$$

To the contrary, in toluene solution no reaction occurs and [Co(Salen)$COOCH_3$] is recovered. These results point to [Co(Salen)$COOCH_3$] being a key intermediate in the reaction scheme with the rate determining step being nucleophilic attack on the methoxycarbonyl moiety by an alcohol molecule. A similar pathway is anticipated for other cobalt complexes.

The stability of the organic ligand under oxidative conditions is a major item for an industrial development of this promising system. Of the cobalt complexes listed in Table VII [Co(Salen)]$_2$·H_2O is the most active, but ligand oxidative degradation stops the catalytic cycle resulting in by- product formation. [Co(pic)$_2$·4H_2O] complex is less active but maintains its activity longer, as observed during a 100 hour test, where only a slight decrease in activity was noted.

Conclusions

The main goals of the chemical industry with respect to environmental issues are a reduction in the use of toxic materials and in the amount of waste generated. The most innovative way to achieve these goals is the development of new process technologies and new products that avoid these environmental concerns. Industrial development of the CuCl catalyzed methanol oxidative carbonylation route to DMC has resulted in DMC playing a significant role in this scenario as a versatile intermediate substituting for dangerous chemicals like phosgene, DMS and methyl chloride, as a non-toxic, non-polluting solvent, and as a component of reformulated fuels.

A recent technology improvement, based on the concept of reactor-evaporator, affords higher productivities in the industrial production of DMC, making it easier to scale up to large capacity plants.

Cobalt complexes appear promising candidates for the development of an alternative non-corrosive catalytic system.

Literature Cited:

1. Massi Mauri, M; Romano, U.; Rivetti, F. *Ing. Chim. Ital.*, **1985**, *21*, 6.
2. Romano, U.; Rivetti, F. XXIst FATIPEC Congress, Amsterdam, NL, 1992, Vol. 1, p. 247.
3. *Hydrocarbon Processing*, **1993**, 2, 23.
4. Bhattacharya, A. K.; Boulanger, E. M. *Preprint Abstract, 208th A.C.S. Nat. Meeting, Div. Env. Chem.* Washington D.C. August 21-25, **1994**, Vol. 34, No. 2 pag. 471
5. Knifton, J.K.; Duranleau, R.G. *J.Mol. Catal.*, **1991**, *67*, 389.
6. Halpern, J. *Comments Inorg. Chem.*, **1981**, *1*, 3.
7. Sonoda, N. *Pure Appl. Chem.*, **1993**, *65*, 699.
8. Delledonne, D.;Romano, U.; Rivetti, F.*U.S. Pat. 5 118 818* (ECS)
9. Fenton, D.M.; Steinwand, P.J. *J. Org. Chem.*, **1974**, *39*, 701.
10. Romano, U.; Rivetti, F. *Chim. Ind. (Milan)*, **1980**, *62*, 7.
11. Romano, U.; Rivetti, F. *J. Organomet. Chem.*, **1979**, *174*, 221.
12. Morris, G.E.; Lucy. R.A. *EP Pat. Appl. 220 863* (British Petroleum)
13. Nisihira, K.; Mizutare, K.; Tanaka, S. *EP Pat. Appl. 425 197*, (UBE)
14. Koch, P.; Cipriani, G.; Perrotti, E. *Gazz. Chim. It.*, **1974**, *104*, 599.
15. Massi Mauri, M; Romano, U.; Tesei, R.; Rebora, P. *Ind. Eng. Chem. Prod. Res. Dev.*, **1980**, *19*, 396.
16. Brubaker, C.H.; Wicholas, M. *J. Inorg. Nucl. Chem.*, **1965**, *27*, 59
17. Rivetti, F.; Romano, U. *U.S. Pat. 5 206 409* (ECS);
18. Rivetti, F.; Romano, U.; *U.S. Pat. 5 159099* (ECS)
19. Romano, U.; Tesei, R.; Cipriani, G.; Micucci, L. *U.S. Pat. 4 218 391*(ECS);
20. Rivetti, F.; Romano, U.; Di Muzio, N. *U.S. Pat. 4 318 862* (ECS)
21. Romano, U.; Tesei, R.; Micucci, L. *Ital. Pat. 1 092 951* (ECS);
22. Curnutt, G. L.; Harley, A. D. In *Procceedings of the fifth IUCCP Symposium, Oxygen Complexes and Oxygen Activation. by metal Complexes*; Editors Martell, A.E. and Sawyer, D.T.; Plenum Press, New York, N.Y., 1987, p. 215.
23. Di Muzio, N.; Rivetti, F.; Fusi, C.; Sasselli, G. *U.S. Pat. 5 210 269* (ECS)
24. Leung, T. W.; Dombek, B.D. *J. Chem. Soc.,Chem Comm.*, **1992**, *3*, 205.
25. Bassoli, A.; Rindone, B.; Tollari, S. *J. Mol. Cat.*, **1990**, *60*, 41.
26. Delledonne, D.; Rivetti, F.; Romano, U. *EP Pat. Appl. 463 678* (ECS).
27. Delledonne, D.; Rivetti, F.; Romano, U. *J. Organomet. Chem.*, **1995**, *488*, C15.
28. Costa, G.; Mestroni, G.; Pellizer, G. *J. Organometal. Chem.*, **1968**, *15*, 187.

RECEIVED November 7, 1995

Chapter 7

Selective Mono-Methylation of Arylacetonitriles and Methyl Arylacetates by Dimethylcarbonate

A Process without Production of Waste

Pietro Tundo, Maurizio Selva, and Carlos A. Marques

Dipartimento di Scienze Ambientali dell'Università di Venezia, Calle Larga S. Marta 2137, Venice 30123, Italy

The reaction of arylacetonitriles and methyl arylacetates with dimethylcarbonate (DMC) at 180-220 °C in the presence of a weak base (K_2CO_3), produces the mono-methylderivatives (2-arylpropionitriles and methyl 2-aryl propionates, respectively) with a selectivity higher than 99%. This reaction has a wide range of applications since the products obtained are well known intermediates for non-steroidal antiinflammatory drugs. Contrary to the usual methylating agents (methylchloride and dimethyl sulfate), the reaction with the non-toxic DMC takes place with only a catalytic amount of the base; accordingly, no inorganic salts are produced. The reaction proceeds both under continuous-flow and batchwise conditions. The mechanism is discussed.

The direct mono-alkylation of methylene-active compounds by common alkylating agents (alkyl halides and dialkyl sulfates) is not a facile reaction and generally cannot be run as a one-step process because considerable amounts of dialkylderivatives form, especially where the methylation reaction is concerned *(1)*. Moreover, the reaction is of a relevant environmental concern; besides the use of highly toxic reagents, alkyl halides and dialkyl sulfates, stoichiometric quantities of inorganic salts are generated as wastes to be disposed of. The methylation of arylacetonitriles $ArCH_2CN$ (**1**) and methyl arylacetates $ArCH_2COOCH_3$ (**2**) has been widely explored; in fact, the corresponding mono-methylated derivatives, 2-arylpropionitriles $ArCH(CH_3)CN$ (**3**) and methyl 2-arylpropionates $ArCH(CH_3)COOCH_3$ (**4**), are intermediates for the synthesis of 2-arylpropionic acids, well-known as non-steroidal antiinflammatory agents *(2)*. Even under phase-transfer catalysis conditions, the direct methylation is poorly selective *(3-4)*. For instance, the reaction of phenylacetonitrile with CH_3I carried out under liquid-liquid phase-transfer catalysis (LL-PTC) produces a mixture

0097–6156/96/0626–0081$12.00/0

of the starting reagent and the mono- and di-methylated derivatives (6, 66 and 28%, respectively) from which the mono-methyl product cannot be separated because of the closeness of the relative boiling points (*5*). Conversely, the use of the non-toxic dimethylcarbonate allows the methylation of both compounds (**1**) and (**2**) to proceed with a selectivity (>99%) not previously observed affording mono-methyl derivatives in a high purity, with no by-products.

The Reactions of Dimethylcarbonate

DMC is an environmentally safe reagent now produced by the oxidative carbonylation of methanol (Equation 1) by Enichem (Italy) (*6*).

$$2CH_3OH + \frac{1}{2} O_2 + CO \xrightarrow{\text{Cu salts}} H_3COCOOCH_3 + H_2O \quad (1)$$

Under batchwise conditions, DMC is a carboxylating agent when reacted with nucleophiles at reflux temperature (90 °C) in the presence of bases (Equation 2) (*7-10*):

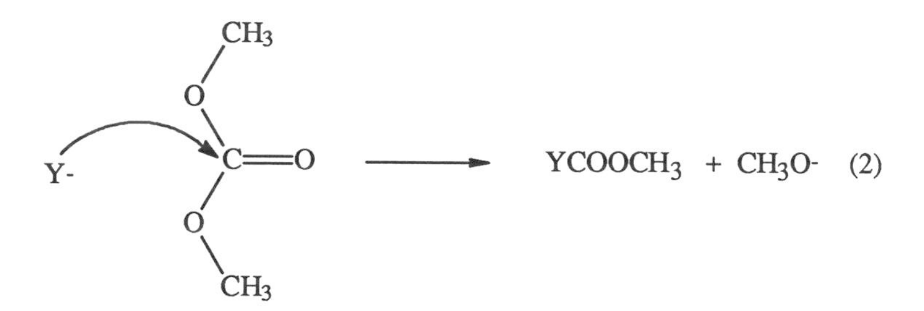

However, operating at high temperatures (≥160 °C), DMC can be used as an excellent methylating agent. The methylation reaction prevails over carboxymethylation since the former is not an equilibrium reaction. In the methylation reaction, the nucleophilic anions attack the methyl group (instead of the acyl carbon) by a $B_{Al}2$ mechanism. The leaving group (methoxycarbonate anion, CH_3OCOO^-) is not stable and rapidly decomposes into methanol and CO_2. Thus, the base can be used in catalytic amounts because the methoxide anion is regenerated (Equation 3).

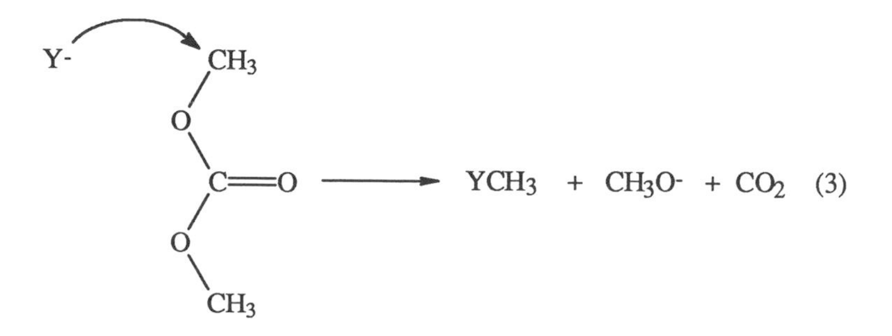

In particular, the methylation reactions by DMC can be run in the presence of weak bases (*e.g.* alkaline carbonates). However, since high reaction temperatures are necessary to exploit the methylating properties of DMC, new methodologies have been developed to perform the reaction under both continuous-flow (c.-f.) and batchwise conditions.

Continuous-Flow Methylations of Dimethylcarbonate. DMC can be used profitably for carrying out methylation reactions under Gas-Liquid Phase-Transfer Catalysis (GL-PTC) conditions (*11*). GL-PTC is a recently reported synthetic method which allows reactions between activated anions and a gaseous organic substrate to be performed under a continuous-flow mode. This technique uses a solid-supported phase-transfer (PT) catalyst which becomes liquid at the reaction temperature; the reactions occur with both the reagents and the products in the gas phase through their continuous transfer between the gas and the catalytic liquid phases. In a typical configuration, a liquid mixture of the reagents is sent continuously to a simple cylindrical catalytic column (plug-flow reactor) where it is vaporized; gaseous products are collected at the outlet of the reactor by a condenser (Figure 1).

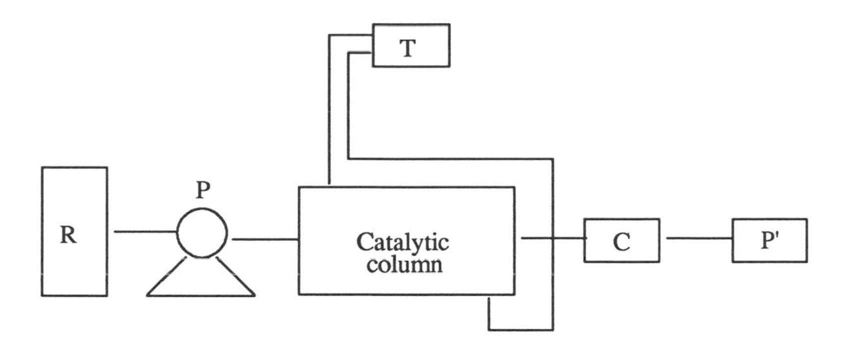

Figure 1. GL-PTC Apparatus: R, reagent's reservoir; P, metering pump; T, thermostat; C, condenser; P', product storage.

When using DMC, polyethylene glycols (PEGs) are the PT-catalysts of choice (*12*); although less efficient than other PT-catalysts (onium salt, crown ethers, cryptands), PEGs are environmentally desirable because they are non-toxic and inexpensive (*13-15*). Under such conditions, in the presence of alkaline carbonates, DMC reacts with methylene-active compounds selectively to give the corresponding mono-methylderivatives. PEGs complex alkaline metal cations so that the basic strength of alkaline carbonates is increased; thus, anionic nucleophiles (conjugate bases of the CH_2-acidic substrates) may be produced in the reaction environment and react with DMC as shown in Equation 3. For example, working at 180 °C and atmospheric pressure, by sending a liquid mixture of DMC and phenylacetonitrile or (*p*-isobutylphenyl)acetonitrile (substrate / DMC in a 1 : 4

molar ratio; flow rate 8 ml/h) on a 100-g catalytic bed composed of 5% PEG 6000 and 5% K_2CO_3 supported on α-Al_2O_3, the corresponding 2-phenylpropionitrile or 2-(*p*-isobutylphenyl)propionitrile are obtained with 99% selectivity at >95% conversion (Equation 4) (*16*).

The 2-(*p*-isobutylphenyl)propionitrile product is an important intermediate; its hydrolysis affords 2-(*p*-isobutylphenyl)propionic acid, better-known commercially as Ibuprofen, the antiinflammatory drug.

Noteworthy is the fact that these reactions produce no hazardous waste. Conversely from the methylations with dimethyl sulfate or methylchloride where a base is also used as a reagent resulting in the generation of stoichiometric quantities of inorganic salts, the reactions of DMC use a catalytic base; CO_2 does not involve disposal problems and the co-product methanol can be easily recycled for DMC production. Actually, these reactions are of a significant environmental interest; they favorably couple safe reaction conditions (c.-f.) to the use of a non-toxic methylating agent in a waste-free process.

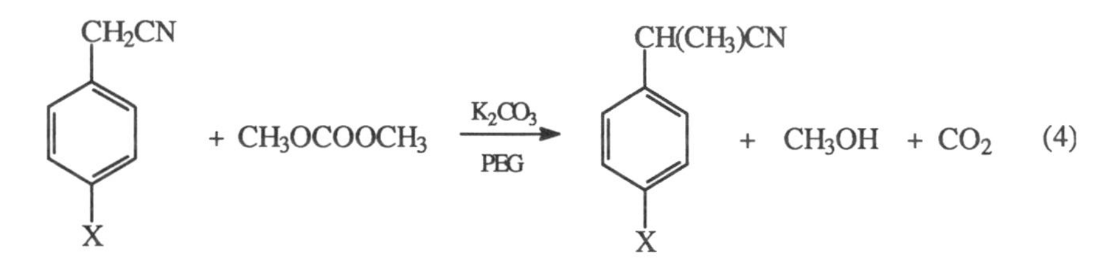

X = H, *p*-isobutyl

Under the same conditions, DMC also reacts with primary aromatic amines and phenols to selectively give mono-N-methylated anilines and anisoles, respectively (*17, 18*).

However, reactions performed with GL-PTC require the reagents to be vaporized and therefore, can be performed only if the compounds have a relatively high vapour pressure. This difficulty can be overcome by reacting DMC under batchwise conditions.

Batchwise Methylations of Dimethylcarbonate. Because DMC boils at 90° C and is a methylating agent only when operating at high temperatures, the batch reactions are necessarily carried out in an autoclave. In a typical experiment, a magnetically stirred mixture of the substrate, DMC and K_2CO_3 (in a 1 : 18 : 2 molar ratio, respectively) is heated in a stainless-steel autoclave, itself heated by an electrical oven. The reaction temperature is checked by a thermocouple dipped into the reaction mixture. Under such conditions, operating at temperatures ranging from 180 to 220 °C (autogenic pressure: 6 to 12 bar), a number of arylacetonitriles and methyl arylacetates have been reacted with DMC; the reaction yielded the corresponding mono-methyl derivatives (2-arylpropionitriles and methyl 2-arylpropionates) with selectivity >99%, (Equation 5 and Table I) (*19-20*) as in GL-PTC reactions.

Table I. Selective Mono-methylation of Arylacetonitriles and Methyl Arylacetates by Dimethylcarbonate[a]

	Substrate $ArCH_2X$	*X*	*Reaction Time / h*	*Conv'n.[b] (%)*	*Product* $ArCH(CH_3)X$ *(% Yield)[c]*
1	Ar = Ph	CN	3.75	100	Ar = Ph (90)
2.	Ar = *o*-$MeOC_6H_5$	CN	14.5	100	Ar = *o*-$MeOC_6H_5$ (85)
3.	Ar = *m*-$MeOC_6H_5$	CN	3.5	100	Ar = *m*-$MeOC_6H_5$ (80)
4.	Ar = *p*-$MeOC_6H_5$	CN	4.75	99	Ar = *p*-$MeOC_6H_5$ (88)
5.	Ar = *o*-MeC_6H_5	CN	7.5	99	Ar = *o*-MeC_6H_5 (82)
6.	Ar = *p*-MeC_6H_5	CN	7.5	98	Ar = *o*-$MeOC_6H_5$ (80)
7.	Ar = *p*-ClC_6H_5	CN	2.25	100	Ar = *p*-ClC_6H_5 (89)
8.	Ar = *p*-FC_6H_5	CN	2.75	100	Ar = *p*-FC_6H_5 (81)
9.	Ar = *m*-$MeO_2CC_6H_5$	CN	8.00	100	Ar = *m*-$MeO_2CC_6H_5$ (91)
10.	Ar = Ph	COOMe	8.00	99	Ar = Ph (80)
11.	Ar = 2-(6-$MeOC_{10}H_6$)	COOMe	6.00	100	Ar = 2-(6-$MeOC_{10}H_6$) (90)

[a]All reactions were carried out in an autoclave and using substrate, DMC and K_2CO_3 in a 1 : 18 : 2 molar ratio, respectively. Entries 1-9: reactions carried out at 180 °C; entries 10-11 reactions carried out at 220 °C. [b]Conversions determined by GC. [c]Yields based on distilled (entries 1-10) or recrystallized (entry 11) products.

$$ArCH_2X + CH_3OCOOCH_3 \xrightarrow{\text{base}} ArCH(CH_3)X + CH_3OH + CO_2 \quad (5)$$

$$X = CN, COOCH_3$$

Two products obtained by this reaction, (*m*-carboxymethylphenyl)-propionitrile and methyl 2-(6-methoxy-2-naphthyl)propionate (entries 9 and 11, Table I), are intermediates for the preparation of the analgesics Ketoprofen and Naproxen (Figure 2), respectively. In particular, the synthesis of Ketoprofen by DMC has been also scaled up in a pilot plant by Tessenderlo Chemie in Belgium (*19*).

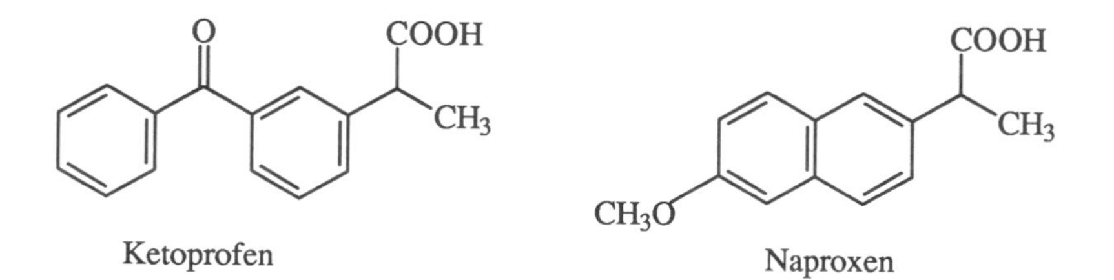

Figure 2. Non-Steroidal Antiinflammatory Agents of the Hydratropic Acids Group.

Batchwise methylation by DMC requires no PT-catalysts, and the base can be employed in a catalytic amount (0.05 mol equivalents with respect to the substrate). Other alkaline carbonates are also effective catalytic bases, their efficacy being clearly related to their solubility in DMC. For example, in using bases from Na_2CO_3 to Cs_2CO_3, the solubility in DMC increases from 0.26 to 0.6 g/L; likewise, the reaction time for the methylation of phenylacetonitrile by DMC at 180°C decreases from 8.75 to 5.75 h, respectively (*20*). The most active base is that in which cation-anion interactions are the poorest and, therefore, a "naked" anion may form (*21*). However, selectivity always remains high. Stronger bases such as phosphazene derivatives (*22*), also promote the reaction but selectivity is lower (Table II).

DMC can be used in a large excess (10-30 molar excess), thus acting both as the methylating agent and the solvent; actually, it has also proved to be the better solvent for these reactions. No improvements in the reaction rate are observed using apolar (cyclohexane), protic polar (methanol) or aprotic polar (DMF) solvent; in particular, the reaction rate is dramatically lowered in cyclohexane solvent (scarce solubility of K_2CO_3), while selectivity is decreased in DMF solvent (Table III).

Table II. Methylation of Phenylacetonitrile with DMC using Different Bases in Catalytic Amounts.[a]

	Base[b]	*Solubility[c] (g/L)*	*Reaction Time /h*	*Conv'n.[d] (%)*	*Selectivity[e] (%)*
1.	Li_2CO_3	0.20	7.25	5	> 99.5
2.	Na_2CO_3	0.26	8.75	89	> 99.5
3.	K_2CO_3	0.58	7.50	98	> 99.5
4.	Cs_2CO_3	0.64	5.75	100	> 99.5
5.	Phosphazene P_1		7.75	99	82.0
6.	Phosphazene P_4		5.0	97	< 1

[a] All reactions were carried out at 180 °C and using $PhCH_2CN$, DMC in a 1 : 18 molar ratio. Bases were used in 0.05 mol equiv. with respect to $PhCH_2CN$. [b]Entries 5-6: *tert*-butyliminotris(dimethylamino)phosphorane and 1-*tert*-butyl-4,4,4-iminotris (dimethylamino)- 2,2,2-bis [tris (dimethylamino) phosphoranylidene amino]-2λ^5,4λ^5-catenadi(phosphazene) were used as bases (Fluka Art. N. 79408 and 79421, respectively); both were completely soluble in DMC. [c]Solubility of alkaline carbonates in DMC was determined by titration (*20*). [d]Conversions determined by GC. [e]Selectivity is defined as [(mol of $PhCH(CH_3)CN$)/(mol of $PhCH(CH_3)CN$ + $PhC(CH_3)_2CN$)] x 100.
Adapted from ref.20.

Table III. Methylation of Phenylacetonitrile with DMC using Different Solvents.[a]

	Solvent[b]	*Reaction Time /h*	*Conv'n.[c] (%)*	*Selectivity[d] (%)*
1.	DMC	3.75	100	> 99.5
2.	Methanol	3.5	96	> 99.5
3.	DMF	3.50	99	90
4.	Cyclohexane	6.0	25	> 99.5

[a] All reactions were carried out at 180 °C and using $PhCH_2CN$, DMC and K_2CO_3 in a 1 : 18 : 2 molar ratio, respectively. [b]A 10 : 1 (v/v) ratio of solvent to substrate was always used. [c] Conversions determined by GC. [d] Selectivity is defined as in Table II. Adapted from ref.20.

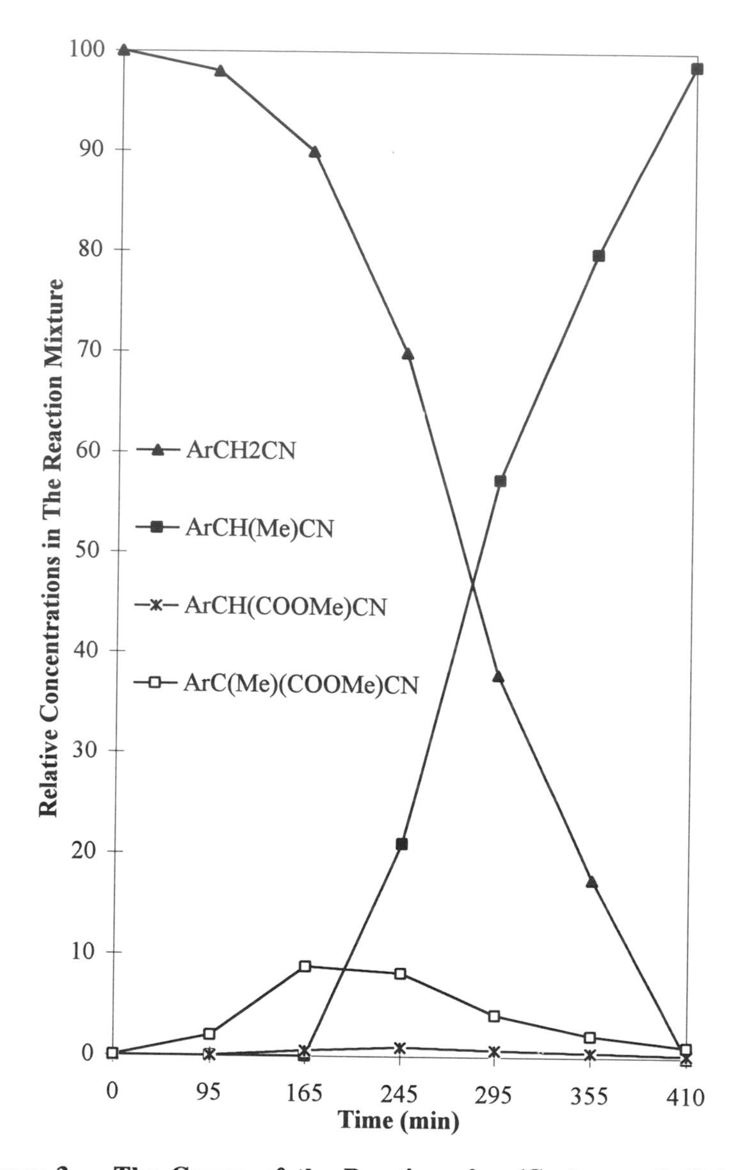

Figure 3. The Course of the Reaction of *m*-(Carboxymethyl)phenylacetonitrile with DMC (Ar = *m*-$MeOOCC_6H_4$) carried out in autoclave at 180 °C (entry 9, Table I). Adapted from ref.20.

The Reaction Mechanism. Two intermediate species, namely ArCH($COOCH_3$)X (**5**) and ArC(CH_3)($COOCH_3$)X (**6**), are detected during the methylation of arylacetonitriles and methyl arylacetates by DMC (*20*). Depending on the reacting substrate, the concentration of (**5**) and (**6**) rises to a maximum (in the range of 6-40%, by GC analysis) and then falls to zero after the complete conversion of the reagent. For example, the course of the reaction yielding (*m*-carboxymethyl phenyl)-propionitrile is depicted in Figure 3. In order to clarify the reaction mechanism, the intermediates (**5**) and (**6**) were synthesized independently from phenylacetonitrile and reacted with DMC at 180 °C in an autoclave. As shown in Figure 4, both compounds (**5**) and (**6**) undergo a reversible decarboxymethylation while the transformation of (**5**) into (**6**) is actually a non-equilibrium methylation reaction which allows the overall reaction (Equation 3) to proceed to completion.

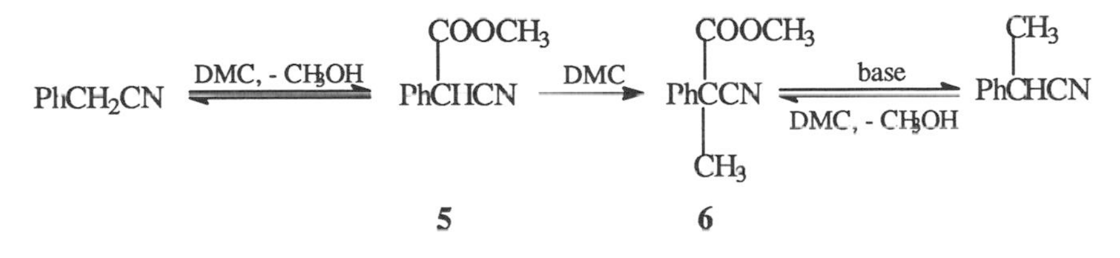

Figure 4. The Reactions of DMC with Compounds 5 and 6 (synthesized from phenylacetonitrile) carried out in autoclave at 180 °C. Substrate, DMC and K_2CO_3 were used in a 1 : 18 : 2 molar ratio, respectively.

The singular selectivity towards the mono-methylation reaction has been explained in terms of two consecutive and very selective reactions involving the two intermediates. That is, *i)* initial attack of an $ArCH^{(-)}X$ (X = CN,$COOCH_3$) anion on the acyl carbon of the dimethylcarbonate ($B_{Ac}2$ mechanism) yielding the carboxymethylated intermediate, ArCH($COOCH_3$)X, is followed by *ii)* attack of the corresponding anion, $ArC^{(-)}$($COOCH_3$)X, on the alkyl carbon of the dimethylcarbonate ($B_{Al}2$ mechanism) yielding the methyl-carboxymethylated intermediate, ArC(CH_3)($COOCH_3$)X. Subsequently, an equilibrium decarboxy alkylation reaction affords the final product, ArCH(CH_3)X. Figure 5 represents the proposed mechanism.

Although the high temperature and the excess of the alkylating agent (DMC) should be expected to decrease selectivity, paradoxically they favor it.

DMC exhibits a double reactivity as a carboxylating and a methylating agent on $ArCH^{(-)}X$ and $ArC^{(-)}$($COOCH_3$)X anions, respectively, and high selectivity actually is a result of the fact that the first anion $ArCH^{(-)}X$ generated attacks only the acyl carbon of DMC and not the methyl carbon. However, the reason for such a behavior is still an open question.

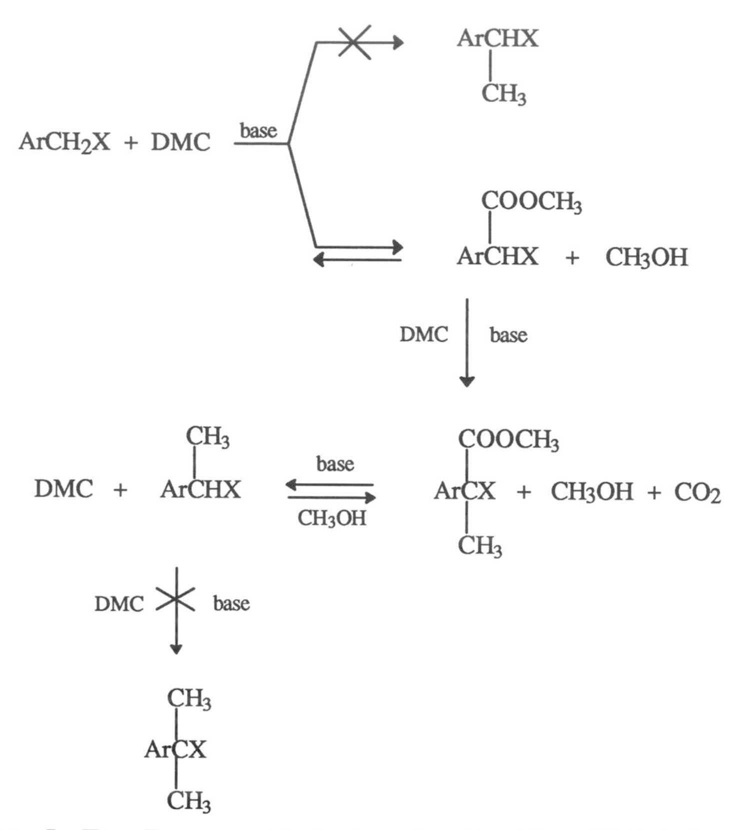

Figure 5. The Proposed Mechanism for the Mono-Methylation of Arylacetonitriles and Methyl Arylacetates.

Conclusions

The mono-methylation of arylacetonitriles and methyl arylacetates with DMC constitutes a new, low environmental impact reaction for the synthesis of high-added-value compounds such as 2-arylpropionic acids. The synthesis has also been successfully scaled up to a pilot plant scale. Besides proceeding in high selectivity (>99%) towards the mono-methyl derivatives, the procedure couples the use of a non-toxic methylating agent (DMC) and a waste-free process in a favorable convergence supporting the need for environmentally cleaner chemistry. The method opens a wide perspective for the development of an intrinsically safer compound (DMC), prepared according to a new technology, to be used either as a methylating agent and/or as an easily biodegradable solvent; also, the absence of chlorinated intermediates (organic or inorganic ones) in the synthesis of DMC means no toxic waste disposal in the overall process here considered. In that sense, the proposed synthetic strategy matches the goal of preventing pollution at the source.

Acknowledgements

This work was supported by the CNR (Consiglio Nazionale delle Ricerche), the MURST (Ministero Universita' e Ricerca Scientifica e Tecnologica) and the Tessenderlo Chemie, Belgium.

Literature Cited

1. Carruthers, W. In *Some Modern Synthetic Methods of Organic Synthesis;* 3rd Ed.; Cambridge University Press: Cambridge, UK, 1989.
2. Rieu, J. P.; Boucherle, A.; Cousse, H.; Mouzin, G. *Tetrahedron* **1986**, *42,* 4095.
3. Starks, C. M.; Liotta, C. In *Phase Transfer Catalysis*; Academic Press Inc.: New York, 1976; ch. 5, pp.170-196.
4. Dehmlov, E. V.; Dehmlov, S. S. In *Phase-Transfer Catalysis*; Verlag Chemie: Veinheim, 1983; ch. 3, pp.123-133.
5. Mikolajczyk, M.; Grzejszczak, S.; Zatorski, A.; Montanari, F.; Cinquini, M. *Tetrahedron Letts* **1975**, 3537.
6. Romano, U.; Rivetti, F.; Di Muzio, N. US Pat. **1979**, 4,318,862; *Chem. Abstr.***1981**, *95*, 80141w.
7. Quesada, M. L.; Schlessinger, R. H. *J. Org. Chem.* **1978**, *43*, 346.
8. *Neth. Appl (ANIC S. p. A.)* 75, 13706, **1976**; *Chem. Abtsr.* **1977**, *87*, 5655.
9. Romano, U.; Fornasari, G.; Di Gioacchino, S. *Ger. Offen.* 3202690, **1982**; *Chem. Abstr.* **1982**, *97*, 144607.
10. Casukhela Someswhara, R.; Prakash Natvarlal, M. *Indian Pat* 141315, **1975**; *Chem. Abstr.* **1980**, *92*, 128583.
11. Tundo, P. In *Continuous Flow Methods in Organic Synthesis*; Horwood, E.: Chichester, UK, 1991.
12. Tundo, P.; Selva, M. *Chemtech* **1995**, *25(5)*, 31-35.
13. Lee, D.; Chang, V. *J. Org. Chem.* **1978**, *43*, 1532.
14. Shirai, M.; Smid, J. *J. Am. Chem. Soc.* **1980**, *102*, 2865.
15. Harris, J. M.; Hudley, N. H.; Shannon, T. G. Struck, E. C. *J. Am. Chem. Soc.* **1982**, *47*, 4789-4791.
16. Tundo, P.; Trotta, F.; Moraglio, G. *J. Chem. Soc., Perkin Trans.* I, **1989**, 1070.
17. Tundo, P.; Trotta, F.; Moraglio, G. *J. Org. Chem.* **1987**, *52*, 1300.
18. Tundo, P.; Trotta, F.; Moraglio, G.; Ligorati, F. *Ind. Eng. Chem. Res.* **1988**, *27*, 1565.
19. Loosen, P.; Tundo, P.; Selva, M. US Patent 5278533, **1994**.
20. Selva, M.; Marques, C.A.; Tundo, P. *J. Chem. Soc., Perkin Trans.* I, **1994**, 1323.
21. Montanari, F.; Landini, D.; Rolla, F. *Top. Curr. Chem.* **1982**, *101*, 147.
22. Schwesinger, R. *Nachr. Chem. Tec. Lab.* **1990**, *38*, 1214.

RECEIVED November 7, 1995

Chapter 8

Oxidation of Phenolic Phenylpropenoids with Dioxygen Using Bis(salicylideneimino)ethylenecobalt(II)

Angela Bassoli[1], Anna Brambilla[1], Ezio Bolzacchini[1], Francesco Chioccara[2], Franca Morazzoni[3], Marco Orlandi[1], and Bruno Rindone[1,4]

[1]Dipartimento di Scienze dell'Ambiente e del Territorio, Università di Milano, Via L. Emanueli 15, 20126 Milan, Italy
[2]Dipartimento di Chimica Organica e Biologica, Università di Napoli, Via Mezzocannone 16, 80134 Naples, Italy
[3]Departimento di Chimica Inorganica, Metallorganica e Analitica, Università di Milano, Via Venezian 21, 20133 Milan, Italy

Recovery of organic carbon from wastewaters deriving from agroindustrial sources is studied using a low-cost catalytic system. This allows both to prevent pollution and to recycle organic materials. The bis(salicylideneimino)ethylenecobalt(II) catalyzed oxygenation of E-ferulic acid in methanol gives vanillic aldehyde, vanillic acid methyl ester and 2-hydroxyhomovanillic aldehyde. Also, methyl 2-hydroxy-3-oxo-(4-hydroxy-3-methoxyphenyl)-propionate is formed from E-methyl ferulate. Electron Paramagnetic Resonance (EPR) measurements show that an organometallic radical is probably the intermediate when the reaction is run in chloroform and in methanol, while the Co(III) superoxo complex is present when the solvent is pyridine. The dependence of the conversion of E-Ferulic acid from the catalytic ratio is also studied. The cobalt-catalyzed oxidation of lignin model compounds, representative of the β-arylether β-O-4) and the phenylcoumaran (β-5) structure, with dioxygen is studied. β-O-4 dimers are cleaved into two fragments, whereas β-5 dimers are oxygenated at the benzylic position with eventual cleavage of the dihydrofuran ring. Pollution prevention benefits could derive from the use of such systems in agroindustrial activities and in wood-processing industry.

Lignin is one of the principal costituents of the woody structure of higher plants, typically 30% of a plant's dry weight, (*1*) and structurally is a highly intricate aromatic polymer of phenylpropenoidic compounds (*2-6*). Oxidative processes are involved both in the polymerization of phenylpropenoidic phenols to lignin and in the degradation of lignin in the enviroment (7-8). The use of model systems for the

[4]Corresponding author

0097–6156/96/0626–0092$12.00/0

study of the competition between oxidative polymerization and oxidative degradation of phenolic phenylpropenoids provides information about the mechanism involved in the biological cycle of lignin (*9*).

Reported here is the use of a Schiff base cobalt complex, bis(salicylideneimino)ethylenecobalt(II), for the oxidation of the phenylpropenoidic phenols, β-aryl ethers and phenylcoumarans, which are used as models of lignin. The β-5 (or phenylcoumaran) linkage accounts for 9-12% of the intermonomer linkages in softwood lignin (*10*). The β-O-4 aryl ethers are the models most commonly employed (*11*) for the study of lignin reactions. These model compounds are also used in degradative studies using fungi such as Phaenerochate chrysosporium, which contains the enzyme ligninase (*12-13*), using microrganisms such as Pseudomona acidovorans (*14*) or mixed bacterial cultures (*15*).

Cobalt (II) complexes with Schiff base ligands are well known to activate dioxygen (*16-18*) and are frequently used to catalyze the oxidation of organic substrates (*19*) and to mimick mono and dioxygenases (*20*). This suggests that the Co-catalyzed oxidation of polyphenols could be a good alternative to the incineration of wastewaters rich in polyphenols. Moreover, the toxicity of Co(II) is much lower than that of other metal ions usually employed as oxidation catalysts.

Results and Discussion

Oxidation of Phenylpropenoidic Monomers. The first group of experiments was devoted to the study of the catalytic oxidation using simple phenylpropenoidic compounds. The first substrate, E-ferulic acid **1**, was oxidized in methanol with dioxygen in the presence of bis(salicylideneimino)ethylenecobalt(II) **2** (Co(II)salen) as the catalyst. The reaction was allowed to proceed for 72 h at room temperature using a substrate:catalyst ratio (R) of 10:1 and 10 bar of dioxygen. The reaction mixture was then methylated with dimethylsulphate and potassium carbonate and separated by alumina chromatography.

Veratraldehyde **3** and veratric acid methyl ester **4** were isolated from vanillic aldehyde **5** and vanillic acid methyl ester **6** in the original reaction mixture. Silica gel chromatography of this reaction mixture was used to isolate small amounts of these primary oxidation products. These compounds had lost two carbon atoms from the original propenoidic chain. A further component isolated from the methylated mixture was 2-methoxyhomoveratric aldehyde **7**, which suggested the presence of 2-hydroxyhomoveratric aldehyde **8** in the oxidation reaction mixture. Its structure was confirmed by independent synthesis. Treatment of the reaction mixture from the catalytic oxidation of E-ferulic acid **1** with 1,3-dithiane followed by methylation formed the thioketal **9**, which was also prepared by reaction of veratraldehyde **3** with 1,3-dithiane carbanion **10** followed by methylation.

A quantitative evaluation of the reaction mixture was obtained by dimethyl sulphate methylation and GC-MS analysis using biphenyl as internal standard. Product yields are shown in Table I.

No reaction was observed in the absence of the catalyst. Moreover, Figure 1 shows that the conversion of E-Ferulic acid **1** was roughly linearly correlated to the concentration of the catalyst.

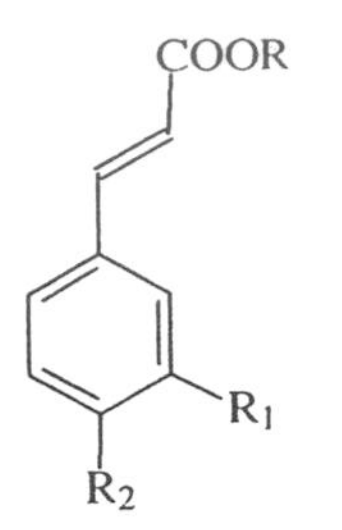

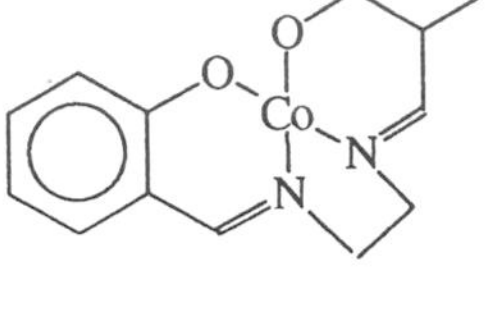

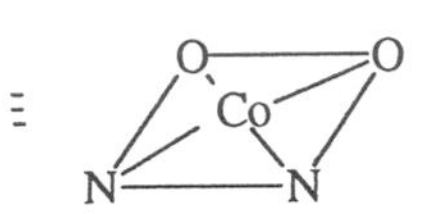

1: R = H; R_1 = OMe; R_2 = OH

11: R = R_1 = OMe; R_2 = OH

13: R = R_1 = R_2 = H

14: R = R_2 = H; R_1 = OH

15: R = H; R_1 = OH; R_2 = OMe

2

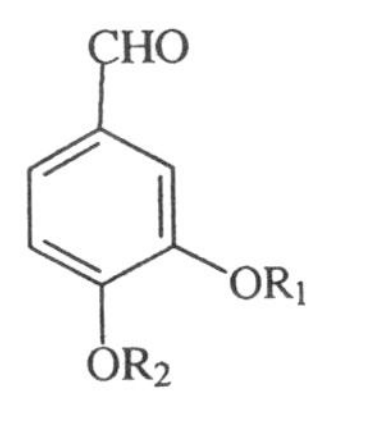

3: R_1 = R_2 = Me

5: R_1 = Me; R_2 = H

COOMe

OR_1

OR_2

4: R_1 = R_2 = Me

6: R_1 = Me; R_2 = H

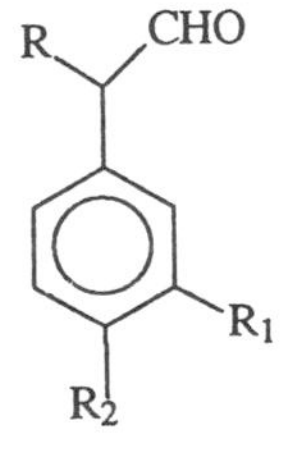

7: R= R_1 = R_2 = OMe

8: R = R_2= OH; R_1 = OMe

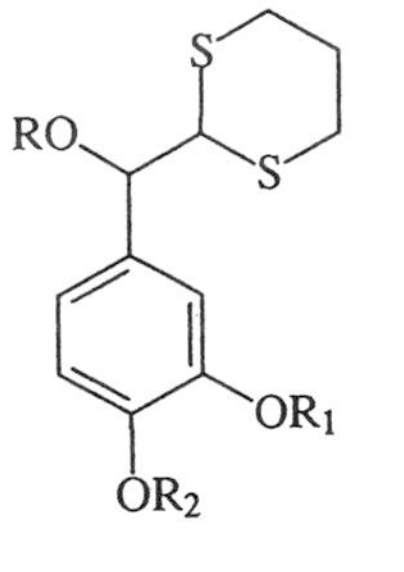

9: R = R_1 = R_2 = Me

S

S

10

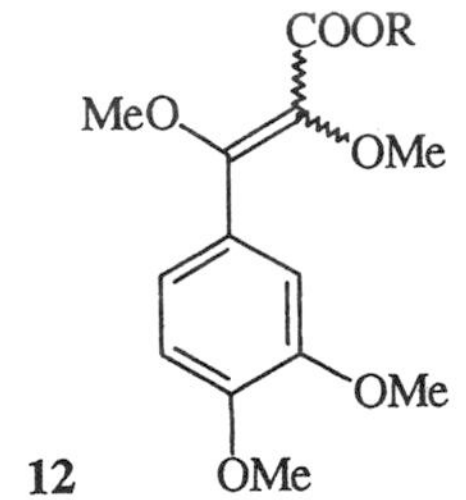

12

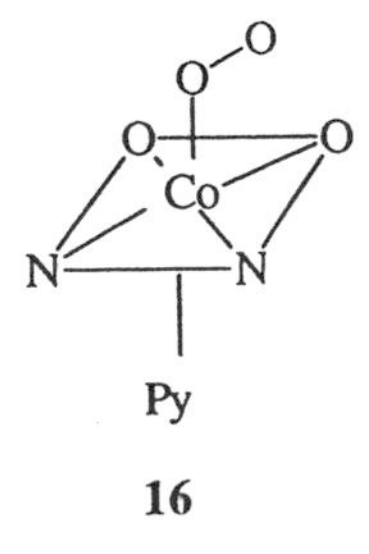

16

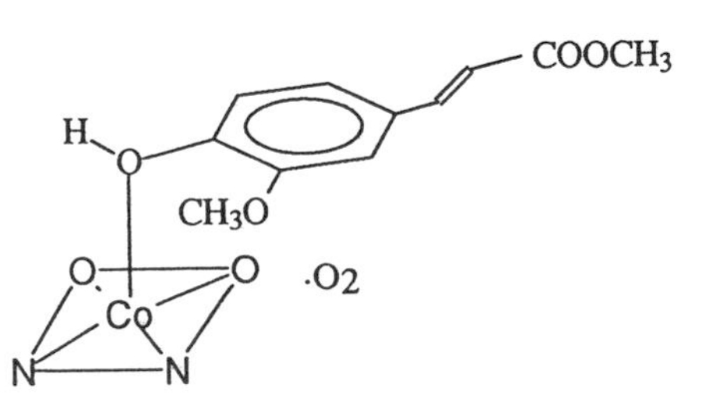

17

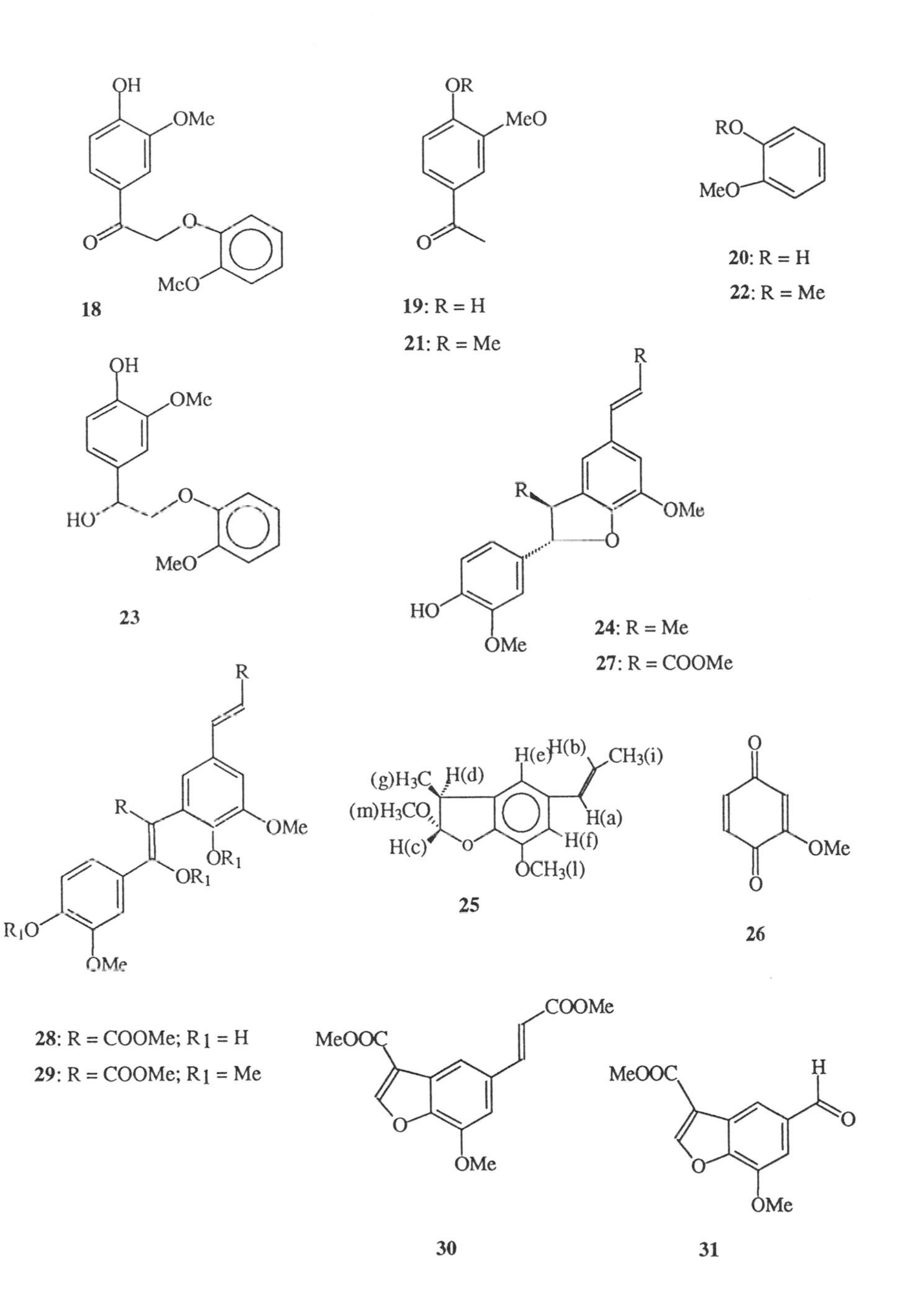

OH
OMe
O
O
MeO
18
OR
MeO
O
19: R = H
21: R = Me
RO
MeO
20: R = H
22: R = Me
OH
OMe
HO
O
MeO
23
R
R
OMe
O
HO
OMe
24: R = Me
27: R = COOMe
R
R
OMe
OR1
OR1
R1O
OMe
28: R = COOMe; R1 = H
29: R = COOMe; R1 = Me
H(e)
H(b)
CH3(i)
(g)H3C
H(d)
(m)H3CO
H(a)
H(c)
O
H(f)
OCH3(l)
25
O
OMe
O
26
COOMe
MeOOC
O
OMe
30
MeOOC
H
O
O
OMe
31

The mode of fragmentation of the original propenoidic chain was further investigated using an ester of ferulic acid, e.g. E-methyl ferulate **11**, as the starting material in the same reaction sequence. The oxidation was repeated at room temperature in methanol using R = 10, $p_{dioxygen}$= 10 bar and t = 72 h; the reaction mixture was methylated with dimethyl sulphate and potassium carbonate and separated by alumina chromatography. Again, veratraldehyde **3** and veratric acid methyl ester **4** were isolated, indicating that the oxidation reaction mixture contained vanillic aldehyde **5** and vanillic acid methyl ester **6**. A new product was also isolated and was determined by GC/MS and HPLC/DAD to be a nearly equimolecular mixture of E- and Z- isomers of 3-(3,4-dimethoxyphenyl)-2,3-dimethoxypropenoic acid methyl ester **12**. Confirmation of the structure came from the ozonolysis to give veratric acid methyl ester **4**. Reaction yields are shown in Table I.

The presence of a hydroxyl group para to the propenoidic chain was necessary for the reaction. In fact, cinnamic acid **13**, 3-hydroxycinnamic acid **14** and 3-hydroxy-4-methoxycinnamic acid **15** did not react under these conditions.

The oxidation of E-methyl ferulate **11** was faster in chloroform than in methanol. In pyridine a very slow reaction was observed (Table II).

Table I. Product Yields % in the Reaction of E-Ferulic Acid 1 and E-Methyl Ferulate 11 with Co(II)salen-dioxygen[a].

Substrate	*St. mat.*	*vanillic aldehyde 5*	*vanillic acid methyl ester 6*	*homovanillic aldehyde 8*	*enediolic ester 12*
E-Ferulic Acid **1**	14	29	41	16	-
E-Methyl Ferulate **11**	11	44	36	-	9

[a]p = 10 bar, R = 10, [S] = 0.01 M, T = 25 °C.

Table II. Conversion vs. Time in the Oxidation of E-Methyl Ferulate 11 with Co(II)salen-dioxygen[a]

time / *solvent*	*conversion 1h*	*conversion 4h*	*conversion 5h*	*conversion 72h*
Pyridine	0%	1%	2%	13%
Methanol	38%	53%	55%	90%
Chloroform	84%	86%	88%	100%

[a]p = 10 bar, R = 10, [S] = 0.01 M, room temperature.

Preliminary EPR measurements during the reaction of E-methyl ferulate **11** shed light on the reaction mechanism. The spectra were performed in the three solvents listed in Table II and showed that the well known Co(II) superoxo complex **16** was present in pyridine, while a different species was formed in chloroform and, in minor amounts, in methanol (*21*). The EPR spectrum of this latter species (Figure 2) had a symmetrical signal constituted from eight lines centered at g = 2.0028. The hyperfine coupling (17 gauss) with the cobalt atom (I = 7/2) suggests that the spin density is mainly located in the organic region of the intermediate. Figures 3, 4 and 5 show the reaction time course of the EPR signal in the three solvents. Structure **17** is tentatively given to the Co(II) radical species present in chloroform and methanol. The amount of the paramagnetic species present in methanol and chloroform with time is in agreement with the conversion per cent (Table II), suggesting that this radical could be the reaction intermediate. The very low conversion occurring in pyridine may be related to the absence of this intermediate.

Oxidation of β-O-4 Models. Lignin is a biopolymer constituted from several types of chemical bonds connecting the monomers. One of the most important of these in the β-O-4 linkage. Two β-O-4 models were used for this experiment. The first β-O-4 dimer reacted with dioxygen (10 bar) in the presence of Co(II) salen as the catalyst was 1-(3,4-dimethoxyphenyl)-(2-methoxyphenoxy)ethane-1-one **18**, prepared according to Landucci (*22*).

After 72 h at room temperature with a substrate: catalyst ratio (R) of 10, the reaction mixture was methylated with dimethylsulphate and potassium carbonate and analyzed by GLC-MS. Veratric acid methyl ester **4** and 1,2-dimethoxybenzene **22** were recognized, indicating that the oxidation reaction mixture contained vanillic acid methyl ester **6** and ortho-methoxyphenol **20**. Also, 3,4-dimethoxyacetophenone **21** was found, indicating the presence of 3-methoxy-4-hydroxyacetophenone **19** in the oxidation reaction mixture.

The second β-O-4 dimer submitted to reaction with dioxygen (10 bar) at room temperature in the presence of Co(II) salen as the catalyst was 1-(3,4-dimethoxyphenyl)-(2-methoxyphenoxy)ethane-1-ol **23**, also prepared according to Landucci (22). In the same reaction conditions, again, veratric acid methyl ester **4** was recognized after methylation, derived from vanillic acid methyl ester **6** present in the oxidation reaction mixture. The other component of the methylated mixture was 1,2-dimethoxy-benzene **22**, derived from ortho-methoxyphenol **20**.

Dimethylsulphate methylation of the reaction mixtures followed by GLC-MS using biphenyl as internal standard provided product yields (Table III).

Oxidation of β-5 Models. Other experiments were devoted to the oxidation of two β-5 dimers. The trans dimer **24** was obtained from E-isoeugenol (*23*). The oxidation with dioxygen (10 bar) in the presence of Co(II) salen as the catalyst gave, after 96 h at room temperature using a substrate:catalyst ratio (R) of 10, a crude reaction mixture which was methylated with dimethylsulphate and potassium carbonate and separated by alumina chromatography. This resulted in the isolation

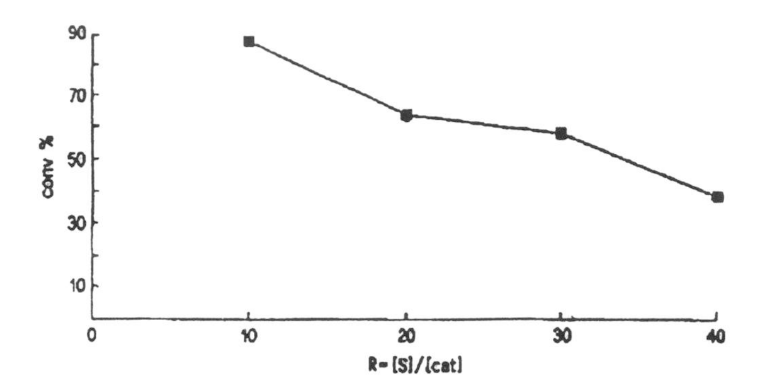

Figure 1. The Dependence of the Conversion of E-Ferulic Acid **1** from the Catalytic Ratio R.

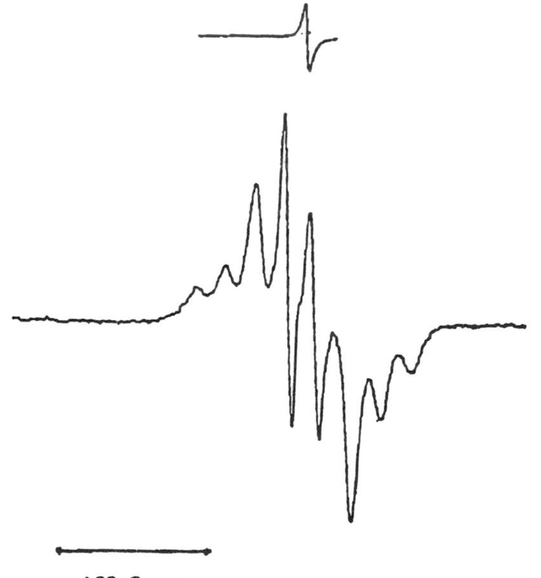

Figure 2. The EPR Signal in the Reaction of E-Methyl ferulate **11** in Chloroform.

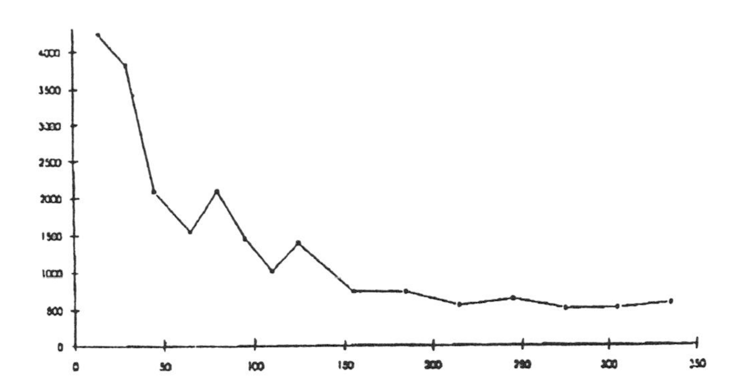

Figure 3. The Time Course of the EPR Signal in the Reaction of E-Methyl Ferulate **11** in Chloroform.

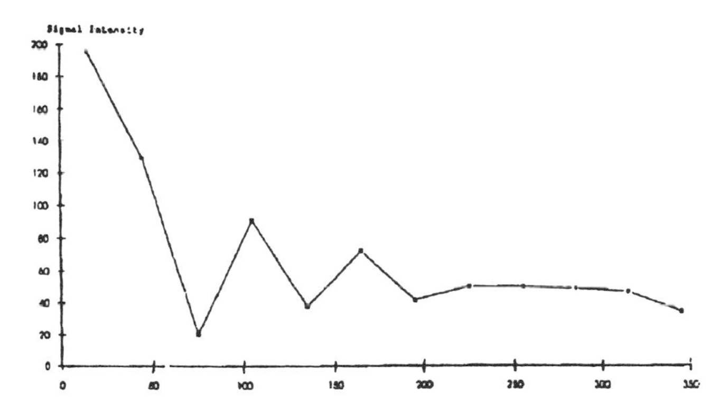

Figure 4. The Time Course of the EPR Signal in the Reaction of E-Methyl Ferulate **11** in Methanol.

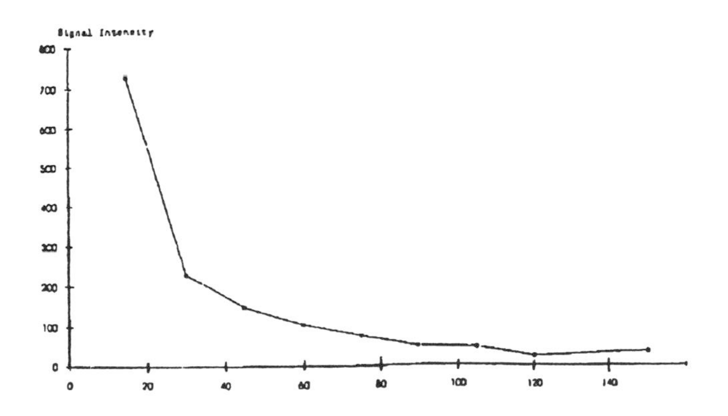

Figure 5. The Time Course of the EPR Signal in the Reaction of E-Methyl Ferulate **11** in Pyridine.

of two fragmentation products: methoxy-para-benzoquinone **26** and the dihydrobenzofuran **25**, characterized by M^+ at m/z = 234 with fragments at m/z = 203, 187. The ^{1}H NMR and the ^{13}C NMR (Table IV) suggest a cis arrangement of H_c and H_d. These stereocenters in the product retain the stereochemistry observed in the starting material **24**.

Table III. Quantitative Results of the Reaction of Lignin Model Compounds with Co(II)salen/dioxygen at 25 °C.

Substrate	*pO2 (bar)*	*R*	*time (hours)*	*Product yields %* Fragmentation	*products*	*Oxygenation products*
18	10	10	72	**21**: 45% **6**:14%	**22**: 14%	
23	10	10	72	**6**: 35%	**22**: 20%	
24	10	10	96	**25**: 25%		
27	10	10	120	**30**: 10%		**29**: 55%
27	1	10	360			**29**: 70%

Table IV. ^{1}H NMR and ^{13}C NMR Spectra of Compound 25.

δ	*^{1}H NMR multiplicity*	*J (Hz)*	*proton(s)*	*^{13}C NMR carbons*	*δ*
1.85	d	6	3H, (g)	C_g	18.2
1.35	d	9	3H, (i)	C_i	18.9
3.41	dq	9, 9	1H, (d)	C_l, C_m	56.5
3.85	s		3H, (l)	C_e	113.9
3,90	s		3H, (m)	C_f	119.8
5,11	d	9	1H, (c)	C_b	124.1
6.15	dq	18, 6	1H, (b)	C_a	131.6
6.32	d	18	1H, (a)		

The second β-5 dimer studied, the trans dimer **27**, was obtained from E-methyl ferulate (*23*). The oxidation with dioxygen (10 bar) in methanol in the presence of Co(II) salen as a catalyst gave, after 120 h at room temperature by using a substrate:catalyst ratio (R) of 10, a crude reaction mixture which was methylated with dimethylsulphate and potassium carbonate and separated by alumina chromatography. Two reaction products were isolated. The most abundant product showed a molecular ion in the mass spectrum at m/z = 472 and fragments at m/z = 457, 442, 428. These data suggested that one oxygen atom had been introduced in the molecule. The ^{1}H-NMR showed seven methoxyl groups between 3.70 and 4.10 δ, as singlets; and a multiplet between 6.80 and 7.20 δ integrating five aromatic protons. The E-propenoidic chain was still present, as indicated by the presence of two doublets at 6.31 and 7.75 δ with a J = 16.6 Hz. The protons of the dihydrofuran ring in compound **27** (two doublets at 6.10 and 6.30 δ) had disappeared. Structure **29** accounted for all these spectral data which suggested that the oxygenation product **28** had been originally formed.

The second component showed a molecular ion in the mass spectrum at 290 m/z, and fragments at m/z = 259, 227, 199. The ^{1}H NMR showed two methoxyl as singlets at 3.83 and 3.92 δ, the E-acrylic side chain consisting of two doublets at 6.45 and 7.75 δ with J = 16.6 Hz, and three aromatic protons as singlets at 7.05, 7.85 and 8.29 δ. This last signal was associated to the aromatic proton nearest to the benzofuran oxygen. This new compound was identified as the benzofuran **30.** Its structure was confirmed by ozonolysis to give the aldehyde **31.**

An optimization of the conversion of compound **27** into compound **29** was obtained using 1 bar of dioxygen. Under these conditions, the yield in compound **29** was 70%.

Table III reports reaction yields for all the lignin model compounds tested.

Conclusions

In conclusion, the Co(II)salen-catalyzed oxidation of β-O-4 and β-5 dimers is a good alternative for the cleavage of carbon-oxygen bonds in lignin models. The cleavage of carbon-carbon bonds is also obtained in some cases. This observation will stimulate further work on the oxidative cleavage of wastewaters containing lignin oligomers for the recovery of organic carbon.

Furthermore, these data provide insight into the efficiency of the catalytic system for the cleavage of lignin model compounds in the scope of the recovery of organic carbon from lignin-containing wastes.

Pollution prevention benefits could derive from the use of such systems in agroindustrial activities and in wood-processing industry.

Experimental Section

General. Column chromatography was performed using alumina Merck 0.07-0.3 mm (R = 100) and eluting with hexane-ethyl acetate 60-20, or using silica gel G Merck (R = 100) eluting with CH_2Cl_2 and a 9:1 mixture of CH_2Cl_2:ethyl acetate. GC/MS was performed using Electron-Impact ionization and a Supelco SPB-5 30 m column with I.D of 0.25 mm and 0.25 μ film thickness. Reverse Phase (RP) HPLC and HPLC-Diode Array Detector (DAD) analyses were performed with a RP-C18 Lichrosorb Merck column. Gradient elution was performed from 60% aqueous acetonitrile to 100% acetonitrile in 20 min at a flow rate of 0.8 ml/min. EPR analyses were performed at 123 K. The values were measured by standardization with DPPH. The ratios between the amounts of radicals were calculated by double integration of the resonance line areas.

Co(II)-catalyzed Reactions. A 10^{-2} M solution of the substrate, 10^{-3} M in bis(salicylideneimino)ethylenecobalt(II) **2** (67 mL) was put in a 100 mL glass liner which was inserted into a 250 mL autoclave. This was then charged with dioxygen (10 bar) and left at 25 °C for the required time. The suspension was then filtered and the solvent evaporated under reduced pressure at room temperature. The residue, was dissolved in 20 mL of acetone, and 0.3 ml of dimethyl sulphate and 435 mg of K_2CO_3 were added. After 2 h at reflux temperature, the solid was filtered and the solution was evaporated under reduced pressure. The residue was then chromatographed over alumina or silica.

Compound **7** had m\z = 210 (M^+), 151; ^1H NMR ($CDCl_3$) δ 3.83, 3.85 and 3.93 (3 s, 3 H each, 3 -OCH_3), 4.61 (d, J = 2 Hz, 1 H, benzylic proton), 6.80-7.00 (m, 3 H, aromatic), 9.59, (d, J = 2 Hz, 1 H, -CHO); IR ($CHCl_3$) 1716 cm^{-1}.

Compound **9** had melting point 89-90 °C. Analytical data calculated for $C_{14}H_{20}O_3S_2$: C 56.00%; H 6.66%; observed: C 55.47%; H 6.36%; MS m/z = 300 (M^+), 268, 181; ^1H NMR ($CDCl_3$) δ 2.05 and 2.85 (m, 6 H, aliphatic protons), 3.85, 3.87 and 3.91 (3 s, 3 H each, 3 -OCH_3), 4.08 (d, J = 7 Hz, -S-CH-S-), 4.83 (dd, 1 H, J = 7, 1 Hz, benzylic proton), 6.8-7.0 (m, 3 H, aromatic); UV (acetonitrile) λ_{max} 285, 235 nm.

The mixture of E- and Z- isomers of 3-(3,4-dimethoxyphenyl)-2,3-dimethoxypropenoic acid methyl ester **12** had m/z = 282(M^+), 267 (M^+-15), 251 (M^+-OMe), 223 (M^+-COOMe); ^1H NMR δ 4.51, 4.56, 4.80, 4.85, 4.90 (5 s, 3 H each, 5 -OCH_3), 6.85-7.25 (m, 3 H, aromatic); λ_{max} (MeOH) 298 nm; IR ($CHCl_3$) 1670 cm^{-1}.

Compound **29** had IR ($CHCl_3$) 2950, 1720, 1460, 1149 cm^{-1}; UV (CH_2Cl_2) λ_{max} 400, 300, 228 nm.

Compound **30** had IR ($CHCl_3$) 3080, 1644, 1461, 973 cm^{-1}; UV (CH_2Cl_2) λ_{max} 230, 265, 289 nm.

Compound **31** had m\z = 234 (M^+), 203; IR ($CHCl_3$) 1734 cm^{-1}; ^1H NMR δ 3.30 (s, 3 H, -OCH_3), 3.95 (s, 3 H, -OCH_3), 7.05 (s, 1 H, aromatic), 7.75 (s, 1 H, aromatic), 7.75 (s, 1 H, aromatic), 9.80 (1H, s, CHO).

Conversion Studies. The appropriate amount of catalyst was added to 50 mL of a 10^{-2} M solution of E-Ferulic acid **1** in methanol and the resulting solution was put in a 250 ml autoclave which was then charged with 10 bar of dioxygen. After 72 h at room temperature, the suspension was filtered and the solvent was evaporated under reduced pressure at room temperature. The residue was dissolved in 10 mL of methanol, methylated with excess ethereal diazomethane, filtered on a silica gel column and analyzed in GC using biphenyl as the internal standard.

Ozonation Reactions. *A* solution of 10 mg of substrate in 70 mL of CH_2Cl_2 was cooled at -10° C, then submitted to a stream of ozone and dioxygen (3.3 g/h of ozone) for 5 seconds. The solution was then treated with a stream of nitrogen, evaporated under reduced pressure and chromagraphed over silica gel.

Treatment of the Methylated Reaction Mixture from E-Ferulic Acid 1 with Propane-1,3-dithiol. The crude reaction mixture was methylated as described above. The mixture, 416 mg, was dissolved in 20 mL of anhydrous benzene, and 0.373 ml (3.73 mmol) of 1,3-propanedithiol and 5 mg (0.026 mmol) of para-toluenesulphonic acid were added. After 2 h reflux in a Dean-Stark instrument (intended to eliminate water during the reaction), the solution was washed with 20 mL water; and the water extract was extracted three times with 20 mL portions of diethyl ether. The combined organic extracts were washed with 20 mL aqueous saturated $NaHCO_3$, dried (Na_2SO_4) and evaporated under reduced pressure. The residue, 300 mg, was then analyzed by RP-HPLC in comparison with an authentic sample of compound **9** using a Diode Array detector.

Synthesis of 1,3-Dithiane-2-[methoxy-(3,4-dimethoxyphenyl)-methane] 9. n Butyl lithium solution in hexane (2 mL, 3.28 mmol) was added to a suspension of 97 mg (0.81 mmol) of 1,3-dithiane in 4 mL of anhydrous THF at -78 °C under N_2. After 2 h of stirring, 123 mg (0.81 mmol) of veratraldehyde in 2 mL of THF was added, and the suspension was kept under the same conditions for 45 min. The temperature was raised to 0 °C, and 0.3 mL (3.15 mmol) of dimethyl sulphate was added. The suspension was refluxed for another 4 h, then 10 mL of water and 10 mL of ethyl acetate were added, and the organic phase was separated. The aqueous phase was extracted with three portions of 15 mL of ethyl acetate,) and the combined organic extractsa were washed with 10 mL of water, dried (Na_2SO_4), and evaporated under reduced pressure to afford a residue which was chromatographed over silica gel.

Acknowledgments

This work was supported by a CNR grant (Piano Finalizzato Chimica Fine II) and by a grant of the Ministero della Pubblica Istruzione. We thank Dr. Monica Casu, Dr. Luisella Bocchio Chiavetto, Dr Suzanne Bach and Dr Valentina Ferrari for technical help.

Literature Cited

1. Prince, R.C.; Stiefel, E.I. *Tibs* **1987**, *12*, 334.
2. Dordick, J. S.; Marletta, M. A.; Klibanow, A. M. *Proc. Natl. Acad. Sci. USA* **1986**, *83*, 6255.
3. Hammel; Kalyanaraman, B.; Kent Kirk, T. *Proc. Natl. Acad. Sci. USA* **1986**, *83*, 3707.
4. Swan, G. A. *Fortschr. Chem. Org. Naturstoffe* **1974**, *31*, 521.

5. Higuchi, T. *Biosynthesis and biodegradation of wood components*; Academic Press: New York, **1985**.
6. Wariishi, H.; Valli, K.; Gold, M. *Biochemistry* **1989,** *28*, 6017;
7. Gold, M. H.; Wariishi, H.; Akileswaran, L.; Mino, Y.; Loher, T. M. *Lignin Enzymic and Microbial Degradation;* Odier, Ed.; INRA Publications: Paris, **1987**.
8. Bassoli, A.; Di Gregorio, G.; Rindone, B.; Tollari, S.; Chioccara, F.; Salmona, M. *Gazz. Chim. Ital.* **1988**, *118*, 763.
9. Basolo, F.; Hoffmann, B. M.; Ibers, J. A. *Acc. Chem. Res.* **1975**, *8*, 384.
10. Adler, E., *Wood Sci. Tech* **1977,** *11*, 169.
11. Nakatsubo, F.; Sato, K.; Higuchi, T. *Holzforschung* **1975,** *29*, 165.
12. Kirk, T.K.; Schultz, E.; Connors, W. J.; Lorenz, L. F.; Zeikus, J. G. *Arch. Microbiol.* **1978,** *277*,1117.
13. Kirk, T.K; Tien, M.; Faison, B. D. *Biotech. Adv.* **1984,** *2*, 183.
14. Vicuna, R.; Gonzalez, B.; Mozuch, M. D.; Kirk, T.K *Applied and Enviromental Microbiology* **1987,** *53*, 2605.
15. Jokela, J.; Pellinen, J.; Salkinoja-Salonen, M.; Brunow, G. *Appl. Microbiol. Biotechnol.* **1985,** *46*, 23.
16. Sheldon, R. A.; Kochi, J. K. *Metal Catalyzed Oxidation of Organic Compounds;* Academic Press, **1981.**
17. Bhatia, B.; Punniyamurthy, T.; Iqbal, J. *J. Org. Chem.* **1993,** *58*, 55.
18. Basolo, F.; Hoffmann, B. M.; Ibers, J. A. *Acc. Chem. Res.* **1975**, *8*, 384.
19. Maddinelli, G.; Nali, M.; Rindone, B.; Tollari, S. *J. Mol Cat.* **1987**, *39*, 71
20. Drago, R. S. *Inorg. Chem.* **1979,** *18*, 1408.
21. Hoffman, B. M.; Diemente, D. L.; Basolo, F. *J. Am. Chem. Soc.* **1970**, *92*, 61.
22. Landucci, L. L.; Geddes, S.; Kirk, T. K. J. *Holzforschung* **1981**, *35*, 66.
23. Chioccara, F.; Poli, S.; Rindone, B.; Pilati, T.; Brunow, G.; Pietikainen, P.; Setala, H. *Acta Chem. Scand.* **1993**, *47*, 610.

RECEIVED December 15, 1995

Chapter 9

Kinetics of Zeolitic Solid Acid-Catalyzed Alkylation of Isobutane with 2-Butene

Michael Simpson, James Wei, and Sankaran Sundaresan

Department of Chemical Engineering, Princeton University, Princeton, NJ 08544

Currently, the alkylation of isobutane with 2-butene is industrially catalyzed using HF or H_2SO_4, which pose environmental and safety threats due to their toxicity and potential for leakage. An environmentally-benign approach to the alkylation of isobutane with 2-butene involves the use of a solid acid catalyst such as the Y-type zeolite. This catalyst, unfortunately, deactivates rapidly during alkylation. A kinetic analysis of the Y zeolite-catalyzed alkylation reaction has been performed to yield insight into how this deactivation may be minimized. The kinetic study is hindered by the rapid catalyst deactivation, requiring an extrapolation of time dependent data to zero-time. The effective reaction rates can be approximated as being first order in butene concentration for a fixed feed isobutane to olefin ratio. When the reactants and products are in the liquid phase, the reaction is severely diffusion-limited for all practical catalyst particle sizes at a temperature of 100 °C.

Alkylation is a very important petroleum refining process that is used to convert isobutane and C_3-C_5 olefins into C_7-C_9 alkanes with high octane numbers. In current industrial practice, this reaction is catalyzed using either hydrofluoric or sulfuric acid. Because of the high toxicity of these acids, safety considerations, and environmental concerns, much effort is being put forward to replace the liquid acid processes with a relatively benign solid acid process. Over the last quarter of a century, a handful of papers have appeared in the open, English literature describing experimental work with solid acid-catalyzed alkylation. It was reported that faujasitic zeolites, in particular Y and rare-earth exchanged X and Y, are initially very active for promoting isobutane/olefin alkylation (*1-5*). The incorporation of rare-earth ions into the zeolitic framework has been demonstrated to improve the catalyst activity, stability, and selectivity (*2*). Recent results have also been reported demonstrating that sulfonated zirconia likewise catalyzes this reaction (*6*). All of these catalysts produce an alkylate rich in trimethylpentanes (TMP's), the C_8's with the highest octane numbers. However, solid acids deactivate rapidly during alkylation. Though the exact nature of the deactivation is not understood, it is believed to be due to the accumulation of a carbonaceous deposit in the micropores

0097–6156/96/0626–0105$12.00/0

of the zeolite (*7*). This deposit may be similar to the conjunct polymers that form during sulfuric acid catalyzed alkylation (*8*). Regardless, it is believed that the cost of this regeneration along with the resulting catalyst degradation make solid acid-based alkylation a relatively uneconomical process compared to the liquid acid-based processes currently employed.

As a first step towards understanding how to minimize the problem of catalyst deactivation during alkylation, an understanding of the fundamental kinetics of alkylation on a solid acid has been sought. Although a number of solid acid catalyzed alkylation studies have presented butene conversion data (*1-6*), none of these have been performed under conditions where a quantitative kinetic analysis is possible. Described in this paper are results from experiments capable of yielding such kinetic measurements. These experiments have been performed in an integral fixed bed reactor loaded with ultra-stable H-Y catalyst. Although a feed I/O ratio in the range of 5 to 15 is used in industrial practice, the catalyst deactivated so rapidly at these compositions that quantitative kinetic analysis was not possible. Therefore, experiments were performed using very low olefin concentrations (I/O ratio in the range of 500 - 3500). Under these conditions the catalyst deactivation occurred over a period of minutes, and it was possible to estimate initial (i.e. zero-time) values characteristic of a clean catalyst. With this information, it will be shown that the alkylation reaction is severely diffusion-limited when using a liquid phase feed.

Experimental

Catalyst. Ultra-stable Y-type faujasite extruded into pellets with approximately 20% inorganic binder was obtained from Union Carbide (LZY-84) and calcined at 400 °C prior to use as the H-form of ultra-stable Y (USHY). The LZY-84 was crushed and sieved to a variety of particle sizes with average diameters, d_p, ranging from 231 to 91 μm, so that the effect of d_p on the catalytic activity could be studied. A framework Si/Al ratio of 6.9 for the LZY-84 was calculated from the unit cell parameter.

Reactor. A steady-flow microreactor constructed entirely of stainless steel was used to measure the conversion and product distribution for the alkylation of isobutane with trans-2-butene. See Figure 1 for a diagram of this reactor. The reaction takes place in a 2" long tube (0.055" ID, 1/8" OD) packed with 2 - 50 mg of catalyst. A high-pressure syringe pump (Isco) was used to deliver a feed mixture of isobutane and trans-2-butene. The mixture was prepared by mixing either pure trans-2-butene (Aldrich) or 5% trans-2-butene in isobutane (Scott Specialty Gases) with isobutane (99+%, Matheson). The isobutane supply contained approximately 4200 ppm of n-butane as a principal impurity. The n-butane was used as an internal standard for gas chromatographic analysis, as was 2,2-dimethylbutane (2,2-DMB, Aldrich). The 2,2-DMB was added during the mixing process in amounts less than 400 ppm and was found to be neither formed nor consumed in this reaction. Though small amounts of n-butane can be formed during alkylation via hydrogen transfer to butene, it is assumed that the amount formed is negligible compared to the quantity in the feed. The n-butane standard was used to normalize the feed injection size, while the 2,2-DMB standard was used to verify the feed concentration of 2-butene based on a known ratio of 2-butene to 2,2-DMB. The feed mixture was dried by passing it over dehydrated alumina prior to loading into the syringe pump. The pump was typically operated at 150-160 cc/hour. The reactor was sustained at pressures higher than 300 psig using a dome-type backpressure regulator (Grove). This pressure was chosen so as to ensure that none of the components in the feed or the effluent would vaporize at any point in the reactor. In-situ activation of the catalysts was achieved by ramping the reactor temperature from ambient to 250 °C at a rate of 2 °C/min and holding at this temperature for about 10 hours. The typical reaction temperature was 100 °C. All heating of the reactor was done using electric heating tape. Ultra-high purity nitrogen (Scott, 99.9995%) was passed through water and oxygen traps and flowed over the catalyst at about 30 cc/min during the activation procedure. The reactor was pre-pressurized with this gas to 300 psig just prior to the beginning of the reaction. Reactor effluent samples were collected directly into vials after flowing through the backpressure regulator and a dry ice-chilled coil. Liquid was collected in this manner over a 5 to 10 second time period. These samples were stored in sealed vials which were immersed in an acetone/dry ice bath. A manual injection into the GC was accomplished by using a chilled and insulated syringe. Typically, about five samples were taken over one minute time intervals. The gas chromatograph (Hewlett-Packard) was equipped with a flame ionization detector and a 100-meter fused silica capillary column specially designed for separating alkylation components (Supelco, Petrocol DH). A fraction of the C_{9+} components did not exit from the capillary column in the given GC run time. The only reactor modification necessary to run vapor phase experiments was to reposition the backpressure regulator upstream of the reactor. This enabled the feed to be delivered at a constant flowrate using the syringe pumps. Since the reactor was at ambient pressure during these experiments, the feed flashed into a vapor after flowing through the backpressure regulator. As with the liquid phase experiments, the effluent was passed through a dry ice bath and sampled frequently.

Results

Alkylate Composition. Table I summarizes the identifiable alkylate product distribution obtained in a representative experiment at different times on stream. In these experiments, the combined selectivity to either trimethylpentanes (TMP) or dimethylhexanes (DMHx) was typically in the 75 - 95% range, and the TMP/DMHx ratio varied from 4.4 to 8.5. The TMP's are the desired products due to their superior octane numbers. As can be seen in table I, a number of C_5-C_7 products were formed in this reaction. These components are likely due to cracking of C_8's or C_{12}'s. The butene isomers, cis-2-butene and 1-butene were also observed as products, though they are not listed in this table. It is thus apparent that both hydrocarbon cracking and butene isomerization occur along with alkylation over the catalyst. It should be noted that the isomerization was not included when determining the rate of reaction. All forms of the butene isomers were treated identically when calculating the extent of conversion.

Extrapolation to Zero-time Kinetics. A striking feature of the alkylation experiments reported in this paper is the steady decline of the butene conversion with increasing time on stream (see Figure 2). Therefore, in order to study the kinetics of alkylation over a fresh catalyst, it is necessary to extrapolate the time-dependent data to zero-time. To facilitate extrapolation, several samples of the reactor effluent were gathered within the first 5 minutes of reactor operation. As can be seen from Figure 2, the rate of deactivation is quite high. Even under dilute 2-butene concentrations, the catalyst deactivates in a matter of minutes. In the remainder of this paper, only the initial (i.e. time zero) conversions will be presented and analyzed. The rate of catalyst deactivation and its relation to operating conditions will be examined in a future publication.

Time-zero Kinetics. At any given particle size and feed concentration, it was found that the dependence of butene conversion on space time can be captured reasonably well by a first order kinetic model. The concentration of isobutane is assumed to be

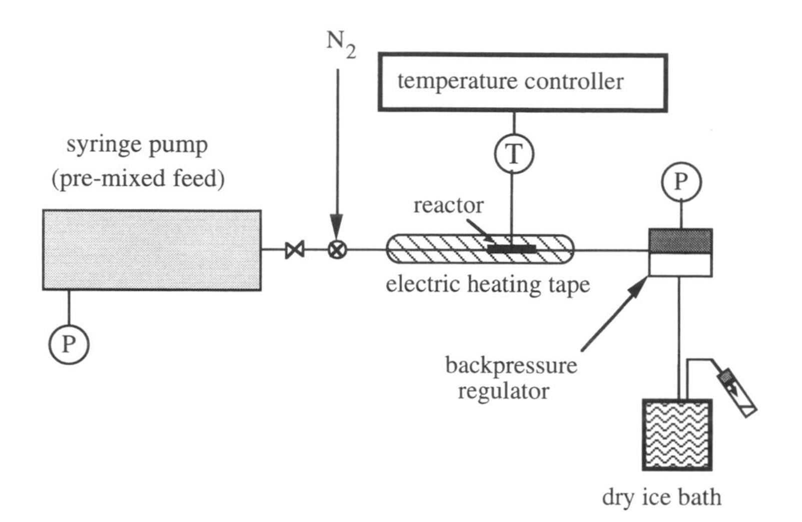

Figure 1. Alkylation reactor system.

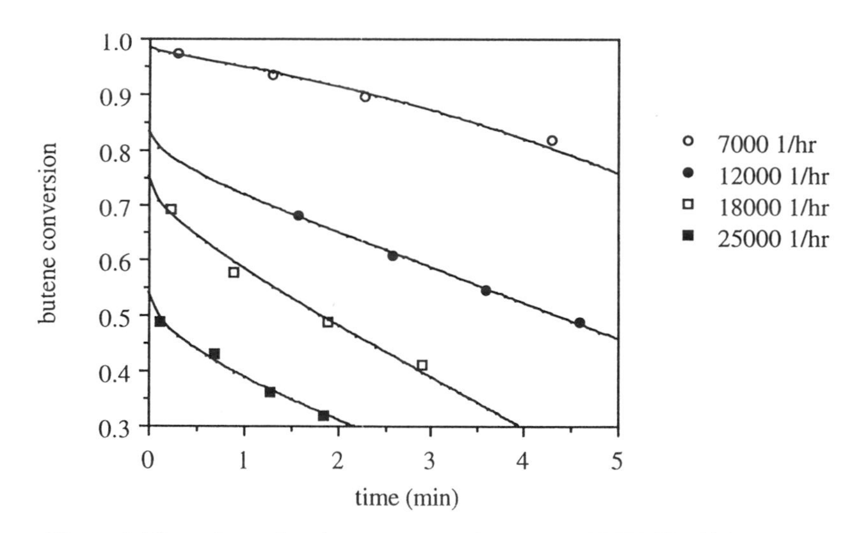

Figure 2. Time-dependent butene conversions over USHY (d_p=231 μm, I/O≈1000, T=100 °C).

constant as it is present in a large excess. Figure 3 demonstrates the consistency between the data and this kinetic model.

Table I. C_{5+} Product distribution for alkylation over USHY (d_p= 231 μm, WHSV= 12000 hr^{-1}, I/O≈1000, T=100 °C).

Product	mole %(0.1 min)	mole %(1.0 min)	mole %(1.6 min)
isopentane	10.0	7.3	7.8
2,3-dimethylbutane	3.9	3.2	3.3
2-methylpentane	0.9	0.7	0.9
3-methylpentane	0.7	0.6	0.7
Total of C_6's	5.5	4.5	4.9
2,4-dimethylpentane	4.7	3.9	4.1
2,3-dimethylpentane	1.7	1.7	1.8
Total of C_7's	6.4	5.6	5.9
2,5-dimethylhexane	0.8	0.8	0.6
2,4-dimethylhexane	5.0	5.2	5.2
2,3-dimethylhexane	3.7	4.1	3.8
3,4-dimethylhexane	0.7	0.9	1.3
Total of DMHx	10.2	11.0	10.9
2,2,4-trimethylpentane	17.8	17.0	16.8
2,3,4-trimethylpentane	25.3	27.7	26.7
2,3,3-trimethylpentane	21.2	23.8	23.2
Total of TMP	64.3	68.5	66.7
Total of C_8's	74.5	79.5	77.6
2,2,5-trimethylhexane	0.0	0.3	0.3
2,6-dimethylheptane	3.6	2.8	3.5
Total of C_9's	3.6	3.1	3.8

It is important to point out that the rate of alkylation is not strictly first-order in butene concentration, since it was found that increasing the I/O ratio increases the fractional conversion. For a first order reaction, the conversion should be independent of the feed concentration. Table II shows the effect of I/O ratio on the conversion for a particular particle size.

Table II. Effect of I/O ratio on the extent of conversion (USHY, d_p=231 μm, 100 °C).

	butene conversion		
I/O ratio	WHSV=12000 hr^{-1}	WHSV=14000 hr^{-1}	WHSV=25000 hr^{-1}
560	0.78	0.78	0.49
820	0.83	0.81	0.54
3350	0.90	0.88	0.63

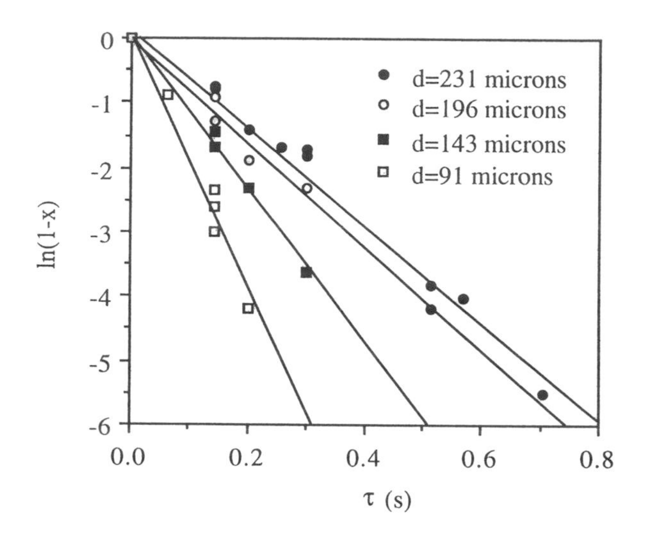

Figure 3. Effect of space time on initial butene conversion over USHY (I/O ≈ 1000, T=100 °C).

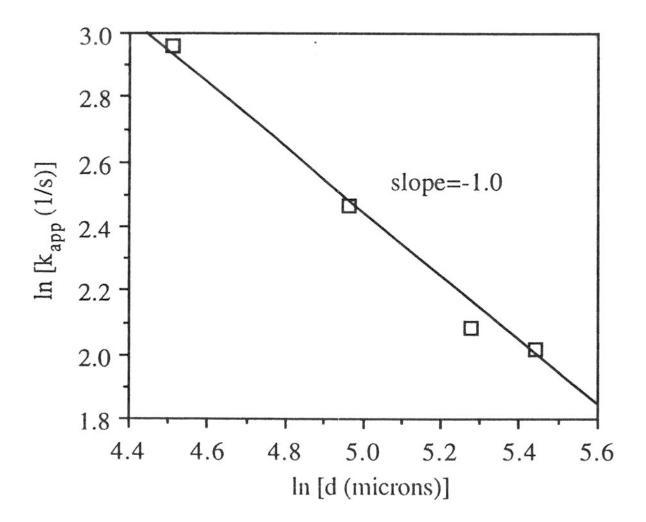

Figure 4. Effect of particle size on the apparent first order rate constant (I/O ≈ 1000, T=100 °C).

Particle Size Effect. For the moment, let it be assumed that the alkylation reaction is approximately first order in butene concentration. It then follows that the apparent order should also be unity in the presence of diffusional limitations. If it is assumed that our alkylation reactor can be modeled as a plug-flow reactor, then the apparent rate constant for each particle size can be readily estimated using equation 1.

$$k_{app} = \frac{1}{\tau} \ln\left(\frac{1}{1 - x}\right) \quad (1)$$

In this formula, τ is defined as the space time. Figure 4 shows a plot of ln k_{app} plotted against ln d_p for experiments at a constant feed concentration and temperature. This result can be correlated as shown in equation 2.

$$\ln k_{app} = 7.7 - \ln d_p$$

i.e.

$$k_{app} = \frac{2200}{d_p} \quad (2)$$

One can readily conclude from equation 2 that the alkylation reaction is not controlled by external mass transfer resistance. If the reaction were limited by external mass transfer resistance, we should have obtained

$$k_{app} \sim \frac{1}{d_p{}^n} \quad (3)$$

where n lies between 1.5 and 2.0.

On the other hand, if the reaction is severely limited by intraparticle diffusion, then

$$k_{app} \sim \frac{1}{d_p} \quad (4)$$

which is indeed observed experimentally. Therefore, it is concluded that all of the data lies in the regime of pronounced intra-particle diffusion control. It can be shown that in this regime

$$k_{app} = \frac{6}{d_p} \sqrt{k\,D} \quad (5)$$

where k is the intrinsic first order rate constant and D is the intraparticle diffusivity.

Another way to demonstrate the influence of intraparticle diffusion is through the well known Weisz-Prater analysis (*9*), which involves calculation of $\eta\ \phi^2$, where

η = effectiveness factor, or

$$\eta = \frac{\text{observed rate per unit volume of particle}}{\text{intrinsic rate per unit volume of particle}} \quad (6)$$

$$\phi = \text{generalized Thiele modulus} = \frac{d_p}{6}\sqrt{\frac{k}{D}} \quad (7)$$

If $\eta\,\phi^2 >> 1$, then the reaction is limited by diffusional resistance. For the present situation,

$$\eta\,\phi^2 = \frac{k_{app}\,d_p^2}{36\,D} \tag{8}$$

Unfortunately, the precise value of the intraparticle diffusivity, D, for butene in the USHY catalyst particles is not known. One should expect D to be smaller than the bulk diffusivity (typically 10^{-5} cm^2/s). It appears reasonable to anticipate D to be in the range of 10^{-7} to 10^{-5} cm^2/s. For d_p=91 µm, the effective rate constant (k_{app}) was found to be 19.3 s^{-1}. The Weisz-Prater numbers can subsequently be calculated for the estimated range of diffusivities.

$$\eta\,\phi^2 = \begin{array}{lll} 4.4 & \text{for} & D = 10^{-5}\ cm^2/s \\ 44 & \text{for} & D = 10^{-6}\ cm^2/s \\ 440 & \text{for} & D = 10^{-7}\ cm^2/s \end{array} \tag{9}$$

Therefore, in spite of the uncertainty in the value of D, one can be confident that the reaction is severely diffusion-limited, and write

$$\eta = \frac{1}{\phi} \qquad \text{so that} \qquad \eta\,\phi^2 = \phi \tag{10}$$

Next, the value of the intrinsic rate constant, k, is estimated from the calculated value of the Thiele modulus.

$$k\,(100\ ^oC) = \begin{array}{lll} 84\ s^{-1} & \text{for} & D = 10^{-5}\ cm^2/s \\ 840\ s^{-1} & \text{for} & D = 10^{-6}\ cm^2/s \\ 8400\ s^{-1} & \text{for} & D = 10^{-7}\ cm^2/s \end{array} \tag{11}$$

These numbers reveal that the (intrinsic) alkylation reaction is exceedingly fast. Using the value of 10^{-6} cm^2/s for the diffusivity, it was found that the generalized Thiele modulus will be unity at 100 ^oC for a particle with a 3.1 µm diameter. This diameter is approximately the same as that of the USHY single crystals! Thus it can be seen that, practically, the reaction can not be run at 100 ^oC in the liquid phase without diffusion limitations masking the intrinsic kinetics.

In current industrial practice, liquid acid catalyzed alkylation of isobutane with trans-2-butene is carried out at high pressure (which drives the reactants and products into a liquid phase) and excess isobutane levels, as both of these factors are known to increase the maximum achievable conversion of the olefin. It has been assumed in all of the previous studies that similar operating conditions would be best for the solid acid-catalyzed alkylation as well. The experiments presented here have also been performed under high pressure conditions. However, the results of this study raise questions on the wisdom of this choice. It is clear from these results that the utilization of the catalyst can be significantly improved by increasing the effective diffusivity of the reactants inside the particles. This suggests that it would be more desirable to operate the alkylation reaction at lower pressures of hydrocarbons where the reactants and products would remain in the gas phase. The diffusivities of the reaction products will also be significantly higher under gas phase conditions which may facilitate their escape from the catalyst particle, thereby lowering the extent of unwanted secondary reactions.

A limited number of gas phase alkylation experiments using USHY as the catalyst have been performed to verify whether or not the catalytic performance is indeed improved by switching from a liquid to a gaseous feed.

Vapor Phase Alkylation. Weisz-Prater analysis, with an assumed value of 10^{-2} cm^2/sec for D, revealed that $\eta\ \phi^2 \approx 0.5$ for vapor phase experiments, which is much smaller than the value calculated for liquid phase experiments. Thus, the role of diffusion is much less significant in vapor phase experiments, which translates to a better utilization of the catalyst. However, the product distribution was adversely affected by the switch to a gaseous feed. Table III gives the product distributions for vapor phase experiments at 100 and 30 °C. These distributions should be compared with the results for a liquid phase experiment at 100 °C given in table I. In switching from a liquid to a gaseous feed, a number of changes are evident. The isopentane selectivity increases by 40%. The C_6 and C_7 selectivities increase 200 to 400%. The C_8 selectivity and TMP/DMHx ratio both decrease by about 50%. Decreasing reaction temperature improves the product distribution by lowering the C_{5-7} selectivities and increasing the TMP/DMHx ratio. However, the isopentane selectivity is still very high relative to the result for the liquid phase experiment.

Table III. Alkylation product distributions for vapor phase feeds over USHY (d_p=143 μm, WHSV=3100-4200 hr^{-1}, I/O≈1000, TOS≈0.3 min).

	mole %	
Product	T_{rxn}=100 °C	T_{rxn}=30 °C
isopentane	14.0	21.3
2,3-dimethylbutane	18.1	8.1
2-methylpentane	5.3	1.9
3-methylpentane	2.8	2.6
Total of C_6's	26.2	12.6
2,4-dimethylpentane	12.4	6.3
2,3-dimethylpentane	4.8	4.5
Total of C_7's	17.2	10.8
2,2,4-trimethylpentane	18.6	14.9
2,5-dimethylhexane	2.1	0.4
2,4-dimethylhexane	5.5	3.6
2,3,4-trimethylpentane	4.4	14.7
2,3,3-trimethylpentane	6.8	19.8
2,3-dimethylhexane	1.6	0.8
3,4-dimethylhexane	0.6	0.4
Total of C_8's	39.6	54.6
2,2,5-trimethylhexane	0.4	0.2
2,6-dimethylheptane	0.0	0.0
Total of C_9's	0.4	0.2
Others	2.6	0.5
TMP/DMHx	3.1	9.6

It is speculated that the intracrystalline I/O ratio may differ for the vapor and liquid phase experiments even when the external I/O ratios are equal. Under vapor phase conditions, the I/O ratio inside the zeolite cavities is likely to be smaller than that in the bulk as a result of preferential sorption of the olefin. Under liquid phase conditions, this difference should be smaller. This may explain the high selectivity towards disproportionation products for the vapor phase experiments. The low intracrystalline I/O ratio faciliates the formation of C_{12}'s, C_{16}'s, etc. which eventually fragment into C_5's, C_6's, and C_7's.

Conclusions

The kinetics of the alkylation of isobutane with 2-butene over the solid acid, USHY, have been investigated. This study was motivated by the need to find an industrially viable solid acid-catalyzed isobutane/butene alkylation system. Currently, this reaction is catalyzed using the liquids, HF or H_2SO_4. Switching from a liquid to a solid acid for this reaction would provide a plethora of environmental benefits. The catalyst toxicity would be virtually nulled., energy-consuming catalyst/alkylate separation processes could be phased out, and the threat of acid leakage would be completely eliminated. Further benefits include the flexibility to use cheaper materials of construction, considering that solid acids are non-corrosive.

An integral fixed bed reactor operating under isothermal conditions and very high I/O ratios was used to analyze the reaction kinetics. The rate of butene conversion varied with time as a result of catalyst deactivation. By extrapolating the data to zero-time, initial butene conversions corresponding to fresh catalysts were estimated. The data on butene conversion obtained at different space velocities suggested that the alkylation reaction is approximately first-order in butene concentration. Varying the I/O feed ratio, however, showed that the first order model is not strictly correct. The conversion is increased by increasing the I/O ratio. An analysis of data obtained with different particle sizes revealed that this reaction is severely diffusion-limited when using a liquid phase feed.

Preliminary vapor phase alkylation experiments have verified that the effect of diffusion is drastically reduced by switching from a liquid to a vapor phase feed. However, the vapor phase alkylation does not appear to be a viable process due to the unfavorable product selectivities observed.

Legend of Symbols

d_p	average diameter of catalyst pellets
D	effective diffusivity of reactants in catalyst pellets
I/O	moles of isobutane per mole of 2-butene
k	intrinsic first order rate constant
k_{app}	apparent first order rate constant
Si/Al	moles of silicon per mole of aluminum in the zeolite framework
TMP/DMHx	moles of trimethylpentane per mole of dimethylhexane
TOS	time on stream
WHSV	weight hourly space velocity
x	time-zero butene conversion
η	effectiveness factor
ϕ	Thiele modulus
τ	space time (mass flowrate of feed/mass of catalyst)

Acknowledgements

We gratefully acknowledge experimental assistance provided by Dr. Weldon Bell of the Mobil Research and Development Corporation (Princeton, New Jersey) and financial support provided by the National Science Foundation (CTS-9216699).

Literature Cited

1. Kirsch, F. W.; Potts, J. D.; Barmby, D. S. *J. Catal.* **1972**, *146*, pp 142-150.
2. Weitkamp, J. In *Catalysis by Zeolites*; Imelik, B., Ed.; ACS Symp. Ser. 55; Elsevier: Amsterdam, **1980**; pp 65-75.
3. Chu, Y. F.; Chester, A. W. *Zeolites* **1986**, *6*, pp 195-200.
4. Khadzhiev, S. N.; Gerzeliev, I. M. *Prepr.-Am. Chem. Soc., Div. Pet. Chem.* **1991**, *36*, pp 799-803.
5. Corma, A.; Martinez, A.; Martinez, C. *J. Catal.* **1994**, 146, pp 185-192.
6. Corma, A.; Martinez, A.; Martinez, C. *J. Catal.* **1994**, 149, pp 52-60.
7. Weitkamp, J.; Maixner, S. *Zeolites* **1987**, *7*, pp 6-8.
8. Albright, L. F.; Spalding, M. A.; Kopser, C. G.; Eckert, R. E. *Ind. Eng. Chem. Res.* **1988**, *27*, pp 386-391.
9. Weisz, P. B.; Prater, C. D. *Adv. Catal.* **1954**, *6*, pp 143-196.

RECEIVED September 21, 1995

Chapter 10

The Role of Catalysts in Environmentally Benign Synthesis of Chemicals

Milagros S. Simmons

Department of Environmental and Industrial Health, University of Michigan, Ann Arbor, MI 48109

Catalysis will play a vital role in the 21st century in the environmentally benign synthesis of new and existing chemicals. These processes will be designed with an emphasize on both environmental and economic factors. This chapter describes the types of catalysts currently used in the syntheses of bulk and fine chemicals and identifies those most likely to have a beneficial impact, economically as well as environmentally, for chemical manufacturing in the future.

In evaluating alternative pathways for the environmentally benign synthesis of chemicals to meet pollution prevention goals, one finds that catalysis can meet all aspects of an effective strategy. In catalyzed reactions, less energy is often required for production and higher efficiency is frequently observed in conversion (resulting in the generation of fewer by-products, co-products, and other potential wastes). In addition, catalysts can be and often are designed to be environmentally safe.

Catalysis and Technology. Catalysis is an important component in a broad range of commercial applications that include large scale processes for petroleum refining and chemical manufacturing. In 1992, worldwide catalyst sales were approximately \$6 billion, of which \$2 billion were US sales *(1)*. More than 60% of the products and 90% of the processes of industry are based on catalysis. Catalysis can also play a critical role in cost-effective, environmentally-compatible technologies. Industry has demonstrated a commitment to developing technologies that are environmentally responsible as well as economically preferable, and catalysis provides an important opportunity to achieve this goal.

0097–6156/96/0626–0116\$12.00/0

Catalysis and Pollution Prevention. Catalysis has already played a significant role in reducing pollution in our environment. This area is well-documented *(2)* by the many applications of catalysts in improving air quality by NO_x removal and emissions control, reducing the use of volatile organic compounds (VOCs), developing alternative catalytic technology to replace the use of chlorine or chlorine-based intermediates in chemical synthesis, and processing and waste minimization.

A continuing role for catalysis in pollution prevention programs is in new synthetic pathways that do not pollute. With catalysis, reactions can be more efficient and selective (thus eliminating large amounts of by-products and other waste components). The challenge for the field of industrial catalysis in the 21st century is to create innovations in catalyst design that will be environmentally responsible, economical, and applicable to industrial syntheses.

This chapter focuses on the ways that catalysts can be used for environmentally benign synthetic processes. Specific illustrations of the different types of catalysts currently used in the syntheses of bulk and fine chemicals are provided, and those most likely to be environmentally and economically beneficial to chemical manufacturing are discussed at greater length. The use of catalysis to reduce energy and material consumption during production is also discussed. These strategies can be considered when designing environmentally benign synthetic pathways and when choosing alternative chemicals that have fewer health and/or environmental concerns than existing chemicals.

Fundamental Research in Catalysis

Although there were considerable advances in the industrial application of catalysis prior to 1990, technological advances generally did not evolve from systematic, fundamental investigations *(3)*. Various interpretations of catalytic phenomena at the surface impeded the progress of understanding the fundamental principles involved.

Historically, researchers developing new catalysts for industrial use have assumed that catalytic phenomena are too complex for fundamental approaches and have instead resorted to trial-and-error. A certain gap still exists between fundamental science and applied technology because most fundamental research is conducted using clean vacuum systems, whereas most practical catalysis is conducted using high-pressure environments. This large disparity in pressure results in a significant difference in the number of molecules striking the catalyst surface, and subsequently, a large difference between the rates of reactions of these systems is observed. Advances in the design and application of catalysts based on fundamental research have been made, however, despite the disparity between the conditions of the two systems.

Types of Catalysts

Homogeneous or heterogenous catalysis can be selected for synthetic processes depending on the number of phases in which the catalytic reaction is conducted.

Homogeneous catalysis is a single-phase reaction, typically liquid/liquid. Reactions are usually stoichiometric with a catalyst turnover rate of one. Homogeneous catalysis has been applied to many processes and continues to be attractive because of the mild conditions that are normally employed. Most industrial applications use organometallic and coordination complexes, although a large number of dissolved homogeneous catalysts are known and are starting to find there way into industrial applications.

Heterogeneous catalysis are bi- or multi-phased; they have dominated the industrial sector even though the fundamental principles involved are largely unknown. Advancements in analytical instrumentation, however, are allowing increased understanding of the catalytic phenomena in these systems. An important aspect of heterogeneous catalysis is the synthesis of active sites via attachment of metal complexes with a given chemical composition to the support surfaces *(4)*.

Metal Catalysts. Transition metals are commonly used as catalysts, particularly in reduction reactions such as hydrogenation. Metal catalysts may be used in bulk form as pure metals, in combination with other metals as bimetallic or multi-metallic mixtures, or may be dispersed on solid supports such as silica, alumina, or carbon *(4)*. The support may influence the reactivity of adsorbed compounds inherent in the structure and morphology of the metal particles dispersed on the support.

An example of how a reaction can be made more environmentally benign through the use of carefully selected metal catalysts is the synthesis of acetaldehyde, shown in Figure 1 *(3)*. The desirable features of this reaction are the use of an aqueous medium, high reaction efficiency, and easy isolation of acetaldehyde as a

$$CH_2{=}CH_2 \quad + \quad 1/2\ O_2 \xrightarrow{\text{catalyst}} CH_3C({=}O)H$$

conventional synthesis: catalyst = $PdCl_2/CuCl_2$
alternative synthesis: catalyst = $PdCl_2$, V-complex

Figure 1. Synthesis of Acetaldehyde

vapor. The conventional reaction, however, requires a large volume of catalyst, typically both $PdCl_2$ and $CuCl_2$ are used. During the course of the reaction, $PdCl_2$ is reduced to Pd^0, $PdCl_2$ is regenerated from Pd^0 by $CuCl_2$, and the resulting $(CuCl_2)^{-1}$ species is oxidized by molecular oxygen, regenerating $CuCl_2$. The large quantity of catalyst required and used results in the generation of a significant amount of chloride ion, shown schematically in Figure 2, which can subsequently react with organic compounds to form chlorinated organic compounds that may have health and/or environmental concerns. Research has recently shown that if $CuCl_2$ is replaced with a vanadium complex stabilized as a heteropolyacid, the amount of $PdCl_2$ used and the amount of chloride ion generated is reduced by up to 100 and 400 times, respectively. In addition, the catalyst is regenerated and has a high turnover rate.

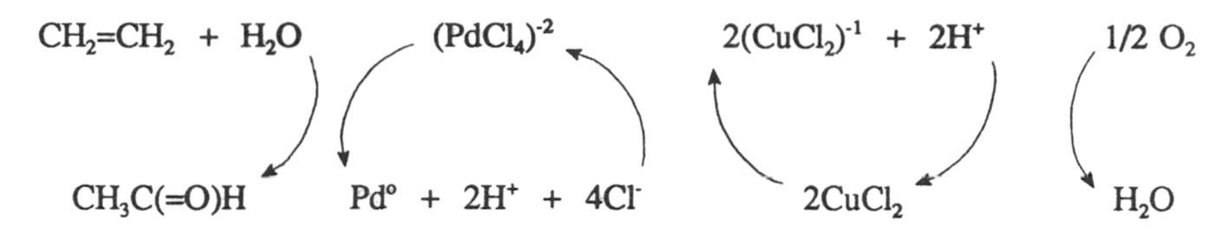

Figure 2. Catalysis of Acetaldehyde

Modifiers can be added to metal catalysts to improve their selectivity and reactivity. These modifiers can be electron acceptors or donors. Alkali metals, for example, are added to increase the bond energy of adsorbed carbon monoxide. Some modifiers enhance selectivity by stabilizing higher oxidation states of metals. Modifiers can also promote desired catalytic reactivity by initiating electronic effects in addition to structural modification of the catalysts. Gold, for example, enhances the catalytic activity of palladium for the oxidation of hydrogen *(5)*. Metal catalysts used in industrial processes have been reviewed by Mouilijn et al. *(6)*.

Metal Oxide Catalysts. Oxides of transition metals have been used for catalytic oxidation. These catalysts may be either non-selective, as in the oxidation of hydrocarbons, or selective, as in the oxidation of olefins using molybdates. Examples of oxidants added in the process are molecular oxygen, hydrogen peroxide, ozone, and other inorganic oxygen donors, such as NaClO, NaBrO, HNO_3, and $KHSO_3$ *(7)*.

In bulk chemical manufacture, the oxidant of choice is molecular oxygen. In the manufacture of fine chemicals, hydrogen peroxide is often preferred although it is more expensive than molecular oxygen. In addition, not only does hydrogen peroxide have 47% active oxygen, but it is also environmentally acceptable since it is converted to water during the oxidation reaction. Ozone is another oxidant of choice that is environmentally acceptable because it is converted to molecular oxygen. Disadvantages to using ozone are that it requires special handling and equipment for its generation.

Inorganic oxygen donors will most likely be replaced in the future by environmentally benign processes; these inorganic donors generate large amounts of inorganic salts when used in chemical syntheses. An example is the synthesis of hydroquinone (Figure 3). In the conventional synthetic pathway, a series of oxidation/reduction reactions generates large amounts of inorganic salts. In the alternative, more benign pathway, the reaction occurs in fewer steps and generates smaller amounts of inorganic salts *(8)*.

Three types of catalytic oxidation processes can occur, each with a unique mechanism *(7)*. These are:

(1) free radical auto-oxidation, in which the metal ion induces the decomposition of alkyl peroxides into radicals via the Haber-Weiss mechanism; this is then followed by the conventional auto-oxidation scheme. Oxidation processes of this type are indiscriminate; hence, they offer poor selectivity.

(2) oxidations of coordinated substrates; these occur when a metal ion oxidizes a coordinated substrate. The reduced form of the metal ion reacts with a terminal oxygen donor to regenerate the oxidized form of the metal ion.

(3) oxygen transfer from an oxygen donor, which could be an oxometal or peroxometal species, to a substrate in the presence of a metal catalyst.

These catalytic oxidation processes are utilized frequently in the manufacture of fine chemicals. Sheldon speculates that future technologies will focus on these processes for the synthesis of chemicals because of their utility and compatibility with environmental mandates *(7)*.

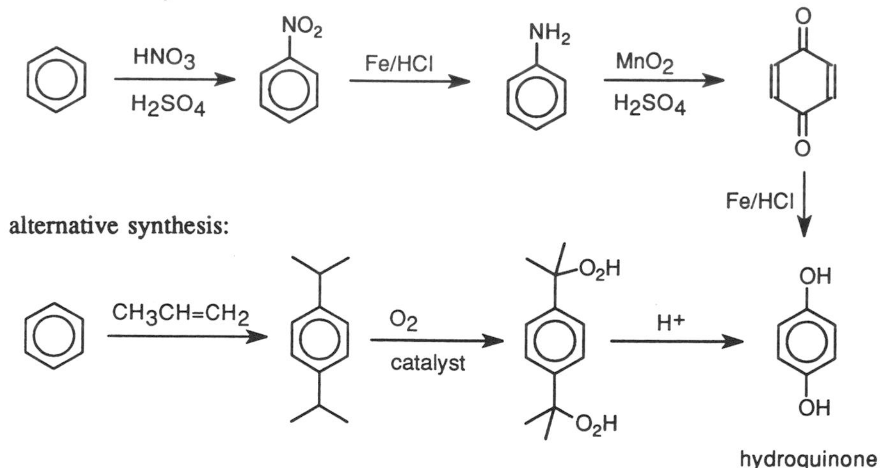

Figure 3. Synthesis of Hydroquinone

Metal Complexes. These catalysts are typically organometallic complexes; they are used predominantly in homogeneous catalysis. Most of these reactions involve oxidative addition of reactants, reductive elimination of products, as well as rearrangements of atoms and chemical bonds in the coordination sphere of the complexed metal atoms *(9)*. An extensive study by Zamaraev on the catalytic property of palladium complexes illustrates the application of these complexes to various chemical syntheses *(10)*.

Chiral metal complexes catalyze reactions in the homogeneous phase and at the same time control the stereospecificity between enantiomers. Selection of proper conditions and the appropriate combination the of central metal and the chiral group(s) are critical in obtaining a high degree of stereospecificity. An important example of this process is the synthesis of the anti-inflammatory drug, Naproxen, shown in Figure 4 *(8)*. Naproxen is obtained quantitatively at 97% yield under high pressure using a transition metal complex containing BINAP [2,2'-bis(diarylphospheno)-1,1'-binaphthyl], an atropisomeric C2 chiral diphosphene in which rotation around the single bond is restricted because of steric factors *(11)*. This type of catalyst is used in the manufacture of pharmaceuticals, flavors, and fragrances. A list of industrial processes using metal chiral complexes is listed in Table I.

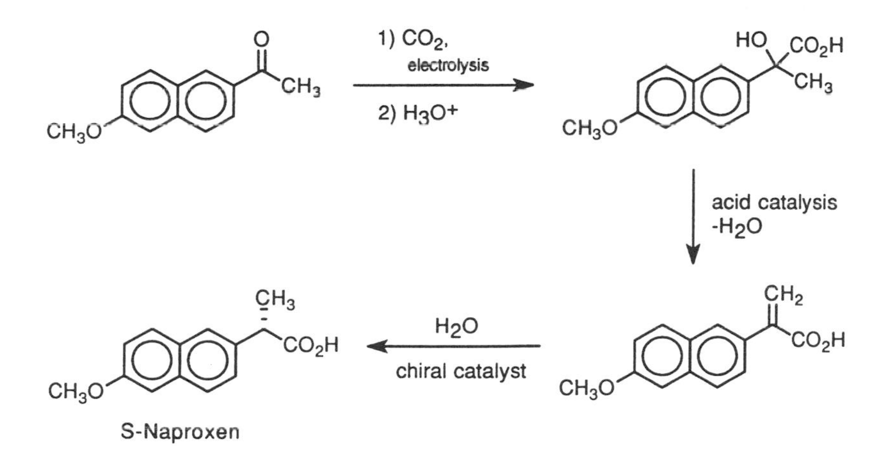

Figure 4. Synthesis of Naproxen

Zeolites. Zeolites are crystalline inorganic polymers made of aluminosilicates and have open framework structures. Natural zeolites (faujasites) have pores of sufficient size to be useful in petroleum refining. Synthetic faujasite-type zeolites are now available in large commercial quantities and have become an important catalyst in the petroleum industry. Comprehensive reviews on the application of zeolite catalysis are available *(12-15)*.

Chemically, zeolites have tetrahedral structures of the type XO_4 in which each X atom is linked by shared oxygen ions. X may be a trivalent (A1, B, Ga), tetravalent (Ge, Si), or pentavalent (P) atom. The size of the zeolite pore openings is determined by the number of tetrahedral units and can be classified as small (6-8 units), medium (10 units) or large (12 units) with a corresponding maximum free diameter of 4.3 angstroms (Å), 6.3 Å, and 7.5 Å. *(14)*.

Table I. Industrial Processes using Metal Chiral Complexes

Process	Catalyst	Reference
Hydrogenation of amides	RuBINAP	Lubell et al., 1991
Isoquinol alkaloid	RuBINAP	Kitomura, 1987
Hydrogenation of terpeneic alcohols	RuBINAP	Takaya et al,1987
Hydrogenation of ketones	RuBINAP	Kitamura et al.,1991
Cyclopropanation of olefins	Cu complex Schiff base	Aratani, T., 1985
Menthol	RhBINAP	Inoue et al.,1990

Zeolites have been used extensively in heterogeneous acid-catalysis involving hydrocarbon transformation. Their high acidity is derived from protons that are required to maintain electrical neutrality, as shown below:

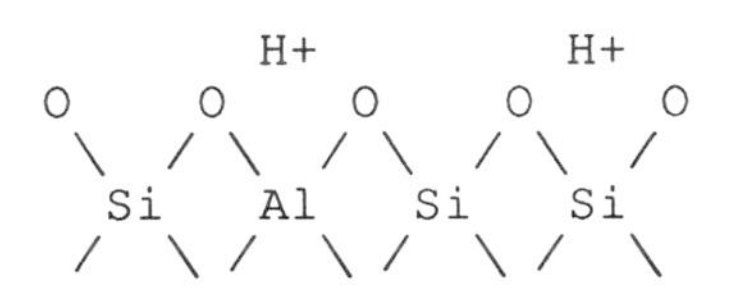

Superacid sites have also been identified by electron paramagnetic spectroscopy suggesting that certain sites possess enhanced acidity due to inductive effects associated with neighboring acid sites *(16)*.

The use of zeolites has also made a dramatic improvement in the performance of fluidized catalytic cracking (FCC) units. An example of a conventional reaction that can be improved with solid acid catalysts such as zeolites is the alkylation of butene (Figure 5) *(3)*. In the conventional reaction, HF or H_2SO_4 are used as the catalyst. Although these reactions are efficient, they use large amounts of corrosive acids and generate large amounts of inorganic salts. HF can be recycled; however, H_2SO_4 needs to be continually removed and replaced. New technologies using solid acid catalysts such as zeolites eliminate the use of corrosive acids and the generation of inorganic salts.

Base-catalyzed and bifunctional (acid/base) applications of zeolites have been reported for bulk chemicals, but they are not as numerous as the acid-catalyzed applications *(15)*. However, base catalysis may play a major role in the synthesis of fine and specialty chemicals. For example, in the synthesis of 4-methylthiazole, a systemic fungicide, basic sites of the Cs zeolite catalyst are responsible for the efficient synthesis of the material without the use of Cl_2, CS_2 or NaOH *(8)*.

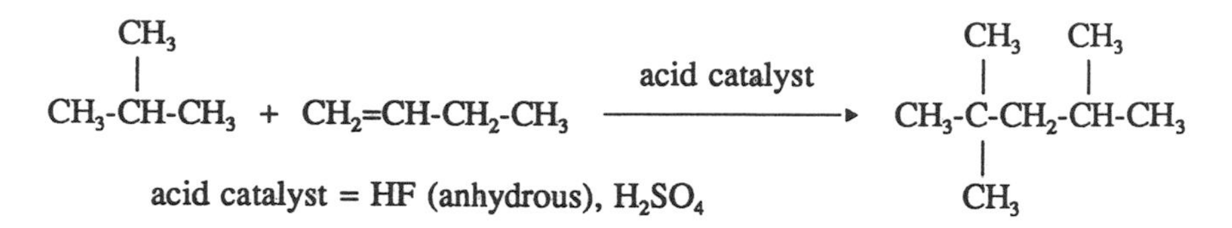

Figure 5. Alkylation of Butene

Chemical modifications of zeolites allow selectivity based on shape making these catalysts quite versatile and promising technology for the future. An example is the synthesis of the 2,6-isomer of diisopropyl naphthalene (Figure 6) *(3)*. The conventional synthesis produces the 2,6- and 2,7-isomers along with the tri- and tetra-isomers. Separation of the 2,6-isomer for use in specialty polymers and liquid crystal polymers is tedious and expensive. The conventional synthesis uses a SiO_2/Al_2O_3 catalyst with a large pore size. The catalyst does not discriminate among the 2,6- and 2,7-isomers and the tri- and tetra-homologs. The use of other zeolites with smaller pore sizes eliminate the production of the tri- and tetra-homologs, but still produce both 2,6- and 2,7-isomers. When using C-mordenite, the 2,6- to 2,7-isomers are produced at a ratio of 7:3 and no tri- or tetra-homologs are formed. These results are shown in Table II.

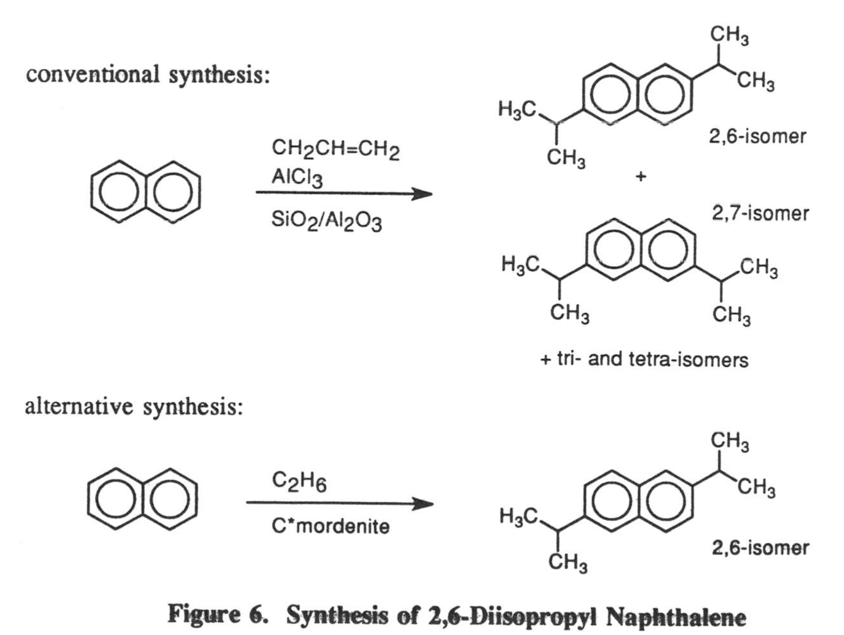

Figure 6. Synthesis of 2,6-Diisopropyl Naphthalene

Table II. Zeolites as Catalysts for the Synthesis of 2,6-Diisopropyl Napthalene

Catalyst	Pore Size (Å)	Ratio of 2,6- to 2,7-isomers	% 2,6-isomer
Si_2O_3/Al_2O_3	60	1	32
Zeolite L	7.1	0.8	22
Zeolite B	7.3	1	37
C*mordenite	7	2.7	70
ZSM-5	5.5	low activity	low activity

There is considerable success in the development of zeolite catalysts, as confirmed by the acceleration in patent applications since the 1960s *(14)*. The pace of development is increasing further as computers are used to design schemes for these catalysts *(3)*.

The potential for zeolites to be used for environmentally benign syntheses is seen in the replacement of substances with obvious health and environmental hazards, such as H_2SO_4 and HF, with zeolites, an innocuous class of substances. In addition to the obvious decrease in toxicity, the increased selectivity and efficiency that zeolites offer serves to enhance their promise in the area of Green Chemistry.

Biocatalysts. In biocatalysis, enzymes and antibodies are used in homogeneous or heterogeneous systems. Reactions that use biocatalysts often proceed with exceptionally high selectivity and as a result, these catalysts are becoming increasingly more important in the asymmetric synthesis of organic compounds.

Enzyme Catalysts. Enzyme catalysts are no longer confined to biochemical reactions in biological systems. They are having an impact on the complex organic syntheses of bioactive molecules and in biotechnology in general. Enzyme catalysts have remarkable properties, and their selectivity makes them desirable and valuable in many synthetic organic pathways. An example is the synthesis of acrylamide (Figure 7).

Enzyme catalysts are regioselective, i.e., they can discriminate among several identical groups within the same molecule. For example, consider the acylation in the pyridine group of castanospermine, an anticancer agent for the treatment of acquired immune deficiency syndrome (AIDS). In this molecule, the four secondary hydroxyl groups of similar reactivity would all undergo acylation under normal conditions with acyl chloride. In contrast, the enzyme catalyst, subtilisin (used in anhydrous pyridine) acylated only the C-1 hydroxyl group *(17)*.

conventional synthesis:

$$CH_2{=}CH\text{-}CN + H_2O \xrightarrow[\text{2. } NH_3]{\text{1. } H_2SO_4} CH_2{=}CH\text{-}C({=}O)\text{-}NH_2 + (NH_4)_2SO_4$$

alternative synthesis:

$$CH_2{=}CH\text{-}CN \xrightarrow[\text{hydratase}]{\text{nitril}} CH_2{=}CH\text{-}C({=}O)\text{-}NH_2$$

Figure 7. Synthesis of Acrylamide

Enzyme catalysts are also chemoselective, i.e., they can select among groups of similar reactivity but of different chemical nature. For example, the catalyst demonstrates a preference toward the acylation of a primary hydroxyl group over a primary amino group during the acylation of adenosine *(18)*.

These catalysts also exhibit stereoselectivity, such as choosing between two stereoisomers of a racemic substrate or between enantiotropic groups in prochiral compounds. A review of enzymes used as catalysts for chiral synthesis has documented the value and utility of these enzyme systems *(19)*.

Enzyme-catalyzed asymmetric syntheses involve two types of reactions: (1) the asymmetric reduction of a prochiral center and (2) the resolution of a racemic material by selective reaction of one enantiomer. Both types are demonstrated in the syntheses of chiral insect phermones reviewed by Sonnet (1988). Enzymes that have broad substrate specificity and still retain other selectivity features can be versatile and powerful catalysts. In addition, enzyme catalysis is applicable not only in aqueous media but also in nonaqueous solvents, including supercritical fluids *(20-22)*. In all cases, however, enzymes require water to function as catalysts. A small amount of water, corresponding to a monolayer on the enzyme molecule, is usually sufficient *(20)*.

Some of the disadvantages to using enzyme catalysts are their high cost, their instability to high temperatures (>250°C), and the difficulty in recovering them for re-use. Interest in enzyme catalysis and technology, however, continues to grow and research is being conducted to alleviate some of these current drawbacks.

Despite the great deal of research that has been conducted on enzymatic catalysis in recent years, the potential of this area of catalysis for use in environmentally benign synthesis has just begun to be realized. Due to both the innocuous nature of enzyme catalysts to human health and the environment and their potential for displacing more hazardous catalysts, the use of enzyme catalysts is in concert with the goals of Green Chemistry.

Antibody catalysts. Another type of biocatalysts is antibody catalysts *(23,24)*. The high specificity and diversity of the immune system are attractive for the production of high-selectivity catalysts from specific binding molecules called antibodies. Antibodies can be created against any molecule with a surface structure. They bind with ligands (haptens or antigens) as do enzymes; these reactions have dissociation constants ranging from 10^{-6} to 10^{-14} *(25)*.

The specificity and selectivity of antibodies correlates with the structure of the antigen used to provide the immune response. These properties can be selected and designed for any given reaction using catalytic antibodies. Hence, tailored antibody catalysts can have far-reaching potentials in organic synthesis and chemical manufacturing.

Early attempts to generate catalytic antibodies were largely unsuccessful *(26)*. However, progress in this field has been steady since 1986 due to the availability of improved transition state mimics and monoclonal antibodies. Lerner, et al. have reviewed chemical reactions amenable to antibody catalysis *(27)*.

Antibodies that bind to metalloporphyrins provide an ideal cofactor in oxidation reactions, similar to cytochrome P-450 catalyzed oxidations. The metalloporphyrins provides the chemical activation site where the antibody exerts its selectivity *(25)*. This antibody-mediated metalloporphyrin chemistry adds another dimension to the use of antibody catalysts in organic syntheses. Research on these systems has demonstrated that antibody technology represents a powerful and versatile tool for creating tailored biocatalysts. Research and development in this area should continue to generate efficient and robust catalysts with high specificity particularly for organic syntheses.

Biomimetic Catalysts. Selective transfers or insertions of atoms such as O and N into various organic molecules under mild conditions are efficiently carried out in biological systems. For example, monooxygenase enzymes catalyze monooxygenation reactions under very mild conditions using cyctochrome P-450, which is ubiquitous in living organisms *(28)*. These types of atom insertions or transfers under mild conditions are challenges in organic chemistry. Research on synthetic catalysts that mimic biological systems has increased over the last decade. This research includes the development of metalloporphyrin oxidation catalysts that mimic cytochrome P-450, such as Fe(TPP)Cl (TPP = tetraphenylporphyrin) using iodosylbenzene as an oxygen donor *(29)*, and selective transfer of nitrogen atoms into hydrocarbons using metalloporphyrins, such as $Fe(TDCPP)ClO_4$ *(30)* and $Mn(TDCPP)ClO_4$ *(31)*.

Some of these biomimetic catalysts have been prepared on inert polymeric and mineral supports *(32)*. These systems are efficient and selective and, at the same time, are practical for preparative and commercial applications. These supported biomimetic catalysts can offer benign synthesis of chemicals, especially if the metals (such as Fe and Mn) can be confined and not leached into the spent materials.

Similar to enzymatic catalysis, biomimetic catalysis offers high selectivity and efficiency. Biomimetic synthesis and the closely related cascade reaction techniques have great potential for accomplishing the goals of Green Chemistry that is beginning to be realized.

Phase-transfer Catalysts. Phase-transfer catalysts (PTCs) promote the reaction between two compounds that are mutually immiscible. Examples are immiscible nonpolar and ionic compounds, for example, alkyl halides and sodium salts, respectively. The phase-transfer agent provides lipophilic cations that transfer the anion of the polar compound from the aqueous to the organic phase as shown below in Figure 8.

aqueous phase:	$Na^+X^- + R_4N^+Cl^-$	$\longleftrightarrow$	$NaCl + R_4N^+X^-$
	$\updownarrow$		$\updownarrow$
organic phase:	$RX + R_4N^+Cl^-$	$\longleftarrow$	$RCl + R_4N^+X^-$

Figure 8. Phase-Transfer Chemistry

PTCs are used for industrial processes and provide increased rates of reaction, higher specificity, and lower energy requirements for manufacturing. They are inexpensive and commercially available. Other cost-effective considerations include the use of inexpensive, non-toxic and recoverable solvents, inexpensive bases for anion generation, and inexpensive oxidants required in the manufacturing process. Some of the commonly used PTCs are quaternary ammonium salts and compounds possessing cation-solvating or binding properties, such as crown ethers and polyethylene glycols. Friedman has reviewed many industrial applications of PTC *(33)*.

Applications of PTCs in organic synthesis include polymer reactions, aromatic substitutions, dehydrohalogenations, oxidations, and alkylations of sugars and carbohydrates. PTCs can be used jointly with organometallic complexes as co-catalysts, bonded to polymeric matrices and used in asymmetric syntheses *(34)*. Industrial applications have been in the manufacture of pharmaceuticals, pesticides, and other chemicals, including epichlorohydrin and benzotrichloride.

One problem associated with phase-transfer catalysis is the large amount of salt generated in the process. This problem was circumvented in a salt-free process developed for the production of aryl ethers using methyl trichloroacetate, which served as a proton sponge in the presence of the co-catalysts, potassium carbonate and 18-crown-6 (see Figure 9) *(35)*.

$$ArOH + RH \xrightarrow[K_2CO_3,\ 18\text{-}C\text{-}6]{CCl_3CO_2CH_3} ArOR + CHCl_3 + CH_3X + CO_2$$

Figure 9. Synthesis of Aryl-Ethers

Future research on PTCs is expected to focus on matrix-supported catalysts adaptable to continuous process as with high resistance to both temperature and strong base. Methods for efficient recovery and trace removal of the spent catalyst will also be developed.

One of the areas under greatest investigation for accomplishing the goals of green chemistry is that of environmentally benign reaction conditions, and in particular, solvent replacement. Through the use of PTCs there is expected to be significant pollution prevention benefits not only through increased efficiency but also in solvent emission reductions.

Aluminum Phosphate-Based Molecular Sieves. Although still in the early stages of development, aluminum phosphate-based molecular sieves have shown great promise as a microporous catalyst and an adsorbent material. They were first reported in 1982 with a neutral aluminum phosphate ($AlPO_4$)-n framework *(15)*. Si, other metals (such as Be, Mg, Ti, Mn, Cr, Fe, Co, and Zn), and other elements (such as Li, B, Ga, Ge, and As) have subsequently been incorporated into the $AlPO_4$ framework; these substitutions have allowed additional applications for the catalyst. Some of the applications reported are catalytic dewaxing, hydrocracking, methanol conversion, and toluene alkylation.

New Catalysts for Future Technology

Catalysis will be critical to the manufacture of chemicals in the 21st century. The design of safer and more efficient catalysts through fundamental scientific research is ongoing and will receive considerable attention in the future. Homogeneous catalysis, which offers high selectivity under mild reaction conditions, will receive considerable attention. A detailed understanding of heterogeneous catalysis at the molecular level will also be an important topic if the goal of selecting suitable and efficient catalysts for environmentally benign synthetic reactions is to be realized.

A current problem associated with heterogeneous catalysis is that high operating temperatures greatly diminish the lifetime of heterogeneous catalysts. High temperatures accelerate deactivation, reduce the selectivity of the catalysts, and hamper their ability to produce high yields. Development of oxidatively and thermally stable ligands and solvent systems for both homogeneous and heterogeneous systems is essential to long catalyst lifetime.

Computer-aided molecular design methods are well suited to modelling reaction pathways, including transition states. These methods can aid significantly not only in the understanding of current catalytic processes but also in the design and identification of new catalysts and even new classes of catalysts.

Catalyst design, process design, and reactor engineering to facilitate separation of products from the catalyst mixture need to be developed concurrently. Innovative approaches, such as using selective membranes for specific catalysts and substrates, are critical in shaping the future of catalysis. Similarly, the development of surface-confined homogeneous catalysts is another valuable approach in the area of process applications. Catalyst supports that offer high site density will provide a basis for further development of highly selective homogeneous catalysts.

Conclusion

Catalyzed reactions are and will continue to play an important role in designing environmentally safer technologies and in the production of safer chemicals. Substitution of environmentally benign synthetic pathways can result in the cost-effective manufacture of chemicals as well as in a large reduction at the source of pollutants in the environment. This chapter has provided an example of how only one functional unit of an entire manufacturing process, that is the use of catalysis, can play an important role in the development of environmentally benign synthesis of chemicals. Options and advances in the field of catalysis that will make a substantial impact in "green" technology were also provided.

Literature Cited

(1) Advanced Technology Program (Department of Commerce) *Catalytica*; submitted.
(2) Armor, J.N. *Applied Catal.* **1992**, *B1*, pp. 221-256.
(3) Cusumano, J.A. In: *Perspective in Catalysis*; Thomas, J.M.; Zamaraev, K.I., Eds.; Blackwell Scientific Publications: Oxford, **1992**, pp. 1-33.
(4) Gates, B.C.; Koningsberg, D.C. *Chemtech*, **1992** (May), pp. 300-307.
(5) Somorjai, G.A. In: *Perspective in Catalysis*; Thomas, J.M.; Zamaraev, K.I., Eds.; Blackwell Scientific Publications: Oxford, **1992**, pp. 147-167.
(6) Mouilijn, J.A.; vanLeeuwen, P.; vanSanten, R.A. In: *Catalysis - An Integrated Approach to Homogeneous, Heterogeneous and Industrial Catalysis*; Mouilijn, J.A.; vanLeeuwen, P.; vanSanten, R.A., Eds.; Elsevier: Amsterdam, **1993**.
(7) Sheldon, R.A. *Chemtech* **1991**, *21*, p. 566.
(8) Dartt, C.B.; Davis, M.E. *Ind. Eng. Chem. Res.*, in press.
(9) Kochi, J.K. *Organometallic Mechanisms and Catalysis*; Academic Press: New York, **1978**.
(10) Zamaraev, K.I. In *Perspectives in Catalysis*; Thomas, J.M.; Zamaraev, K.I., Eds.; Blackwell Scientific Publications: Oxford, **1992**.
(11) Noyori, R. *Chemtech* **1992**, pp. 360-367.
(12) *Zeolite chemistry and catalysis*; Rabo, J.A., Ed.; Monograph #171; American Chemical Society: Washington, D.C., **1976**.
(13) Dwyer, J. *Chem. and Industry* **1984** (April 2), pp. 258-269.
(14) Chen, N.Y.; Degnan, T.F. *Chem. Eng. Progress* **1988** (February), pp. 32-41.
(15) John, C.S.; Clark, D.M.; Maxwell, I.E. In: *Perspective in Catalysis*; Thomas, J.; Zamaraev, J., Eds.; Blackwell Scientific Publications: Oxford, **1992**, pp. 387-430.
(16) Lunsford, J.H. *J.Phys.Chem.* **1968**, *72*, pp. 4163-4168.
(17) Margolin, A.L.; Delinick, D.L.; Whalon, M.R. *J. Am. Chem. Soc.* **1990**, *112*, pp. 2849-2859.
(18) Riva, S.; Chopineau, J.; Kieboom, A.P.G.; Klibanov, A.M. *J. Am. Chem. Soc.* **1988**, *110*, p. 584.
(19) Jones, J.B. *Tetrahedron*, **1986**, *42*, pp. 3351-3363.

(20) Klibanov, A.M. *Chemtech* **1986**, *16*, pp. 354-359.
(21) Dordick, J.S. *Enzyme Microbiol. Technol.* **1989**, *11*, pp. 194-211.
(22) Aaltonen, O.; Rantakyla, M. *Chemtech* **1991** (April), pp. 240-248.
(23) Schultz, P.G. *Science*, **1988**, *240*, pp. 426-433.
(24) Janda, K.D.; Benkov, S.J.; Lerner, R.A. *Science* **1989**, *244*, pp. 437-440.
(25) Keinan, E.; Sinha, S.C.; Sinha-Bagchi; Benory, E.; Ghozi, M.C.; Eshhar, Z.; Green, B.S. *Pure and Appl. Chem.* **1990**, *62*, pp. 2013-2019.
(26) Raso, V.; Stollar, B.D. *Biochemistry* **1975**, *14*, pp. 591-599.
(27) Lerner, R.A.; Benkovic, S.J.; Schultz, P.G. *Science* **1991**, *252*, pp. 659-667.
(28) Ortiz de Montellano, P.R. *Cytochrome P-450: Structure, Mechanism and Biochemistry*; Plenum Press: New York, **1986**.
(29) Groves, J.T.; Nemo, T.E.; Myers, R.S. *J. Am. Chem. Soc.* **1979**, *101*, pp. 1032-1033.
(30) Mahy, J.P.; Bedi, G.; Battioni, P.; Mansuy, D. *J. Chem. Soc., Perkin Trans. II* **1988**, pp. 1517-1524.
(31) Mahy, J.P.; Bedi, G.; Battioni, P.; Mansuy, D. *Tetrahedron Letters* **1988**, *29*, pp. 1927-1930.
(32) Battioni, P.; Lallier, J.P.; Barloy, L.; Mansuy *J. Chem. Soc., Chem. Commun.* **1989**, pp. 1149-1151.
(33) Freedman, H.H. *Pure and Appl. Chem.* **1986**, *58*, pp. 857-868.
(34) Goldberg, Y. *Phase Transfer Catalysis: Selected Problems and Applications*; Gordon and Breach Science Publishers: Amsterdam, **1992**.
(35) Renga, J.M.; Wang, P.C. *Synth. Commun.* **1984**, *14(1)*, pp. 69-76.

RECEIVED December 29, 1995

ALTERNATIVE REACTION CONDITIONS

Chapter 11

Supercritical Carbon Dioxide as a Substitute Solvent for Chemical Synthesis and Catalysis

David A. Morgenstern[1], Richard M. LeLacheur[1], David K. Morita[1], Samuel L. Borkowsky[1], Shaoguang Feng[1], Geoffrey H. Brown[1], Li Luan[1], Michael F. Gross[2], Mark J. Burk[2], and William Tumas[1,3]

[1]Chemical Science and Technology Division, Los Alamos National Laboratory, Mail Stop J514, Los Alamos, NM 87545
[2]Chemistry Department, Duke University, Durham, NC 27706

Supercritical carbon dioxide represents an inexpensive, environmentally benign alternative to conventional solvents for chemical synthesis. In this chapter, we delineate the range of reactions for which supercritical CO_2 represents a potentially viable replacement solvent based on solubility considerations and describe the reactors and associated equipment used to explore catalytic and other synthetic reactions in this medium. Three examples of homogeneous catalytic reactions in supercritical CO_2 are presented: the copolymerization of CO_2 with epoxides, ruthenium-mediated phase transfer oxidation of olefins in a supercritical CO_2/aqueous system, and the catalyic asymmetric hydrogenation of enamides. The first two classes of reactions proceed in supercritical CO_2, but no improvement in reactivity over conventional solvents was observed. Hydrogenation reactions, however, exhibit enantioselectivities superior to conventional solvents for several substrates.

The use of supercritical fluids, particularly carbon dioxide, as a substitute solvent for chemical synthesis is an area of rapidly growing importance. (*1-5*) Carbon dioxide, when compressed to a liquid or, above its critical point (T_c = 31.1°C, P_c = 1071 psi, ρ_c = 0.468 g/ml), to a supercritical fluid (SCF), represents an environmentally benign alternative to organic solvents. Since CO_2 is nontoxic, nonflammable, inexpensive, and unregulated it can replace hazardous organic solvents and thereby provide a valuable pollution prevention tool. Moreover, as discussed below, there is a significant potential for improved synthetic chemistry via faster rates and/or enhanced selectivity for a number of reactions.

An advantage of supercritical fluids as reaction media is that several important solvent properties can be "pressure-tuned" from gas-like to liquid-like, providing opportunities for better chemistry. Supercritical fluids can have liquid-like densities and solvent strength (*vide infra*), so they can be good solvents. Furthermore, the solvent strength of SCF's can be "tuned" by adjusting the density of the medium (and therefore the density-dependent solvent properties such as dielectric, viscosity etc.) via pressure control, potentially leading to more controllable and selective reactions. (*3,6-8*) Supercritical fluids, however, also share many of the advantages of gases

[3]Corresponding author

0097–6156/96/0626–0132$12.00/0

including lower viscosities, (*9,10*) higher gas miscibilities and greater diffusivities, (*11-15*) thereby providing the potential for faster reactions, particularly for diffusion-controlled reactions or processes involving gaseous reagents such as hydrogen, oxygen, or carbon monoxide. A further advantage of CO_2 is that it cannot be oxidized further and thus could be an ideal solvent for carrying out oxidation reactions. Lastly, one should be able to capitalize on the high concentration of CO_2 and use it as a reagent in addition to a solvent, potentially leading to higher rates or even new chemistry. (*16*)

It is probably fair to state that synthetic chemists are relative latecomers to the field of supercritical fluid science and technology. In contrast, the physical chemistry of supercritical fluids has been intensively investigated over the last two decades. There has been considerable progress in areas such as critical phenomena, (*17*) solvation and solubility behavior, (*18-20*) and solvent-solute clustering. (*21,22*) Three applications of SCF's have developed to a commercial or near-commercial level: supercritical fluid extraction and chromatography, including the commercial extraction of caffeine from coffee using supercritical CO_2, (*23-28*) precision cleaning using CO_2, (*29-31*) and waste treatment in supercritical water ("hydrothermal processing"). (*32,33*) The technological and scientific advances that have emerged from these areas provides a strong framework for identifying potentially significant processes for chemical synthesis using supercritical CO_2.

Recent work by several research groups has shown that supercritical fluids can be superior to other solvents for several chemical processes. For example, DeSimone has demonstrated the ability of supercritical CO_2 to replace Freons in the free radical polymerization of fluorinated acrylate monomers. (*34*) Noyori has shown that significant rate enhancements can be achieved in supercritical carbon dioxide relative to other solvents for the homogeneous catalytic hydrogenation of carbon dioxide to either formic acid or its derivatives in the presence of triethylamine or triethylamine/methanol respectively. (equation 1). (*35-37*) As discussed below, we have recently demonstrated that improved enantioselectivities can be achieved in supercritical carbon dioxide for the catalytic asymmetric hydrogenation of several enamides. (*5,38*)

$$CO_2 + H_2 \xrightarrow[\text{SC } CO_2]{Ru(PMe_3)_4H_2} HCO_2H \xrightarrow[NEt_3]{} (HNEt_3^+)(HCO_2^-)$$
$$CO_2 + H_2 \xrightarrow[\text{SC } CO_2]{Ru(PMe_3)_4Cl_2} HCO_2H \xrightarrow[NEt_3]{MeOH} HCO_2Me + H_2O \qquad (1)$$

A number of other reactions have been reported in supercritical carbon dioxide, and the area has been the subject of several recent reviews. (*1-4*) Carbon dioxide has already proven to be a workable, and in some cases superior solvent for polymerizations, (*34,39-43*) electrophilic reactions, (*44,45*) enzymatic transformations, (*2,46-49*) and other reactions. (*50-52*) Homogeneous catalysis constitutes a particularly promising area in which the potential for selectivity gains and improved catalyst separation adds to the economic incentive for adopting supercritical CO_2-based synthetic methods. Areas investigated to date include synthesis of formic acid, (*35-37*) asymmetric hydrogenation, (*38*) and hydroformylation of olefins. (*53*)

Despite these demonstrations of the advantages of carrying out chemical reactions in supercritical fluids and the recent surge in interest in this area, entire classes of catalytic and other synthetic processes have yet to be explored. In this chapter, we describe the approach that our research group has used to explore chemical synthesis and catalysis in supercritical carbon dioxide. We have found that the methodology described below allows for investigation of a wide range of reactions in supercritical CO_2 almost as easily as one would change conventional solvents. We also discuss the limitations of supercritical CO_2 as a solvent and the range of reactants and reaction conditions which can be readily employed. Consideration of the solvent properties of supercritical CO_2, both qualitatively and using the Hildebrand picture allows an easy initial evaluation of the types of reactions and substrates that should be effective in this medium. We then outline our criteria for identifying promising reactions to test and present three examples from work in our laboratory at Los Alamos: catalytic asymmetric hydrogenation of enamides, two-phase ruthenium-catalyzed oxidation of cyclohexene, and copolymerization of CO_2 with epoxides.

Carbon Dioxide as a Solvent for Synthesis

1) Qualitative Considerations. The most important constraints on synthetic chemistry are those imposed by the solvation properties of supercritical CO_2. The solubility of a wide variety of organic compounds has been determined in both subcritical and supercritical CO_2. (*54*) Solubility in SCF's is a strong function of density, a fact which is exploited in supercritical fluid chromatography to improve resolution via pressure programming, (*23-28,55*) and can be dramatically enhanced by the use of organic cosolvents. (*56,57*) Although solubility behavior in supercritical CO2 is a complex subject which is not yet fully understood, several generalizations relevant to synthetic chemistry can be made. The regime of interest to the synthetic chemist interested in pollution prevention is likely to be that which provides maximum solubility (densities above 0.8 g/ml) while using minimal cosolvent (solvent replacement). In this regime, the solubility properties of supercritical CO_2 are similar to those of hexane, although CO_2 has some hydrogen bond acceptor capabilities as well as some dipole selectivity.(*58*) Thus, nonpolar organic species below molecular weight 400 are generally highly soluble, including alkanes, alkenes, aromatics, ketones, and alcohols. Highly polar compounds such as sugars and amino acids are generally quite insoluble. Polysiloxanes and fluoropolymers are quite soluble, (*59*) but most other polymers are not. Limitations on polymer solubility can sometimes be overcome via the use of phase stabilizers or emulsifiers, which typically contain a "CO_2-philic" fluorinated component along with a hydrocarbon portion. (*42,60*)

Two other limitations on the solubility of simple compounds exist. First, salts are generally not soluble, ruling out most ionic reactions and the use of most ionic catalysts in supercritical CO_2. Several strategies employing chelators, phase transfer agents or highly lipophilic counterions have been used to introduce ionic species into supercritical CO_2. Metal cations have been extracted into supercritical CO_2 by converting them to neutral complexes via in-situ chelation using fluorinated ligand precursors such as FOD (FOD = 1,1,1,2,2,3,3-heptafluoro-7,7-dimethyl-4,6-octanedione) (equation 2). (*61,62*) Phase transfer catalysis has been reported in CO_2 using

$$M^{n+} + \text{FOD} \longrightarrow M(FOD)_n + nH^+ \qquad (2)$$

M = Lanthanide FOD ($CF_2CF_2CF_3$)

quaternary ammonium salts and acetone cosolvent (5.2% by volume). (*63*) We have recently reported that the use of fluorinated anions such as "BARF" (BARF =

tetrakis(3,5-*bis*(trifluoromethyl)-phenyl)borate) can lend sufficient solubility to cationic transition metal complexes to enable them to be used for catalytic reactions in supercritical CO_2 even without cosolvents. (*38*)

Another limitation stems from the fact that CO_2 is an electrophile and can undergo specific interactions with Lewis bases. The most important example is the reaction of primary and secondary alkyl amines with CO_2 to form carbamic acids (equation 3). This reaction is well known in conventional solvents (*64*) and has been used as a route to isocyanates and urethanes. The carbamic acid reacts with another equivalent of amine to form an insoluble salt. (*65,66*)

$$RNH_2 + CO_2 \longrightarrow RN(H)C(=O)OH \xrightarrow{RNH_2} RNH_3^+ + RN(H)C(=O)O^- \quad (3)$$

Some free amine may remain in equilibrium with the salt. Tertiary amines lack the tautomerizable proton required for this reaction and are thus soluble. In addition, several workers have noted that reaction 3 does not occur for less basic amines such as aniline which are soluble in supercritical CO_2. (*67*) These results suggest the existence of a basicity limit for solubility of amines (and possibly other compounds) in supercritical (and subcritical) CO_2 which is thought to be roughly $pK_b > 9$. (*68*)

It follows that carbon dioxide can be expected to be a useful substitute solvent for reactions of neutral, relatively volatile compounds using neutral catalysts. Moreover, syntheses involving ions and other less soluble species are sometimes possible via the use of lipophilic cations, phase transfer agents, emulsifiers and other methods. In practice, high solubilities are likely to require high densities. Table I, which summarizes most synthetic reactions reported to date in supercritical CO_2,

Table I: Conditions for Reactions in Supercritical CO_2

Reaction	P (psi)	T (°C)	ρ (g/mL)	HSP*	Reference
Hydroformylation	3000, 5000	80	0.60, 0.78	5.15, 6.69	53
Bromination	2500, 5000	40	0.82, 0.94	6.98, 8.01	52
Organometallic synthesis	3000	ambient	0.95	8.08	51
Polymer synthesis	3000, 5000	40-65	0.70-0.94	6.01-8.01	34,42
Polymer synthesis	5000	60	0.87	7.38	43
Hydrogenation	3000	50, 100	0.49, 0.79	4.16, 6.80	35-37
Asymmetric hydrogenation	5000	40	0.94	8.01	38

* Hildebrand Solubility Parameter. Densities calculated using the Pitzer equation (*69,70*) as implemented in the program SFSolver (Isco Inc., Lincoln, NE).

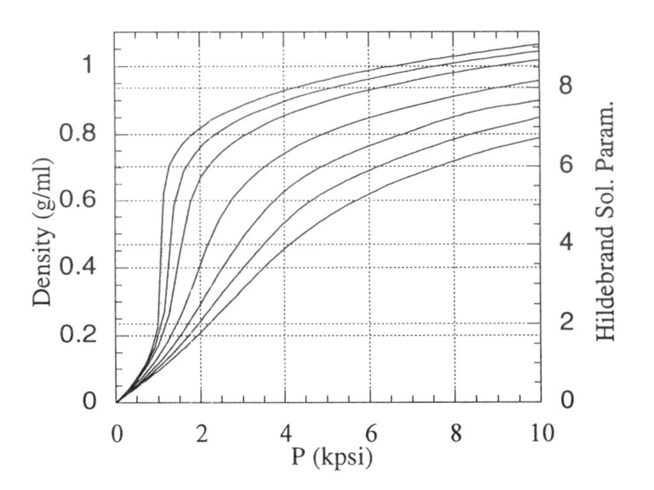

Figure 1: Plot of density and Hildebrand solubility parameter for neat supercritical CO_2 vs. pressure at various temperatures. Hildebrand solubility parameters calculated using equation 6 From top to bottom, the curves represent temperatures of 32.2°C, 40°C, 50°C, 75°C, 100°C, 125°C, and 150°C.

confirms this prediction. Reactions conducted below 60°C have, in practice, virtually always been conducted at densities of 0.8 g/ml or higher. Reactions at higher temperatures involve a compromise between the need for higher densities and the experimental inconvenience of operating at significantly higher pressures.

2) Pressure/Temperature Effects and Limitations on Solubility. A more quantitative approach to selecting a pressure and temperature for reactions in supercritical carbon dioxide can be taken by considering the Hildebrand solubility parameter, δ, of supercritical CO_2 as a function of temperature and pressure. Solvent parameters have proven particularly important in the area of supercritical fluid extraction and chromatography. For example, the direct replacement of Soxhlet extraction by supercritical CO_2 at a pressure and temperature which produce a Hildebrand solubility parameter identical to the Soxhlet solvent has been shown to lead to identical chromatograms for extraction of fiber coatings. (*71*)

The Hildebrand solubility parameter for conventional liquids, defined in equation 4, is largely a measure of the attractive interaction between liquid molecules.

$$\delta = \left(\frac{{}_{1}\Delta_{g}U}{V} \right)^{1/2} \qquad (4)$$

In equation 4, ${}_{1}\Delta_{g}U$ represents the energy required to vaporize a molecule in a molar volume of solvent V. δ has units of $(\text{calories/cm}^3)^{1/2}$. Giddings and coworkers have shown that if the van der Waals equation of state is used for liquids, δ becomes a function of the intermolecular attraction parameter, a, as in equation 5, which Giddings has cast in a form that depends on the critical parameters of the solvent, (equation 6). (*72*)

$$\delta = \frac{a^{1/2}}{V} \qquad (5),$$

$$\delta = (3P_c)^{1/2}\rho_r \qquad (6).$$

In equation 6, P_c is the critical pressure of the solvent and ρ_r is the reduced density (ratio of the density to the density at the critical point). Thus, for a van der Waals gas, the solubility parameter is a function only of density. An equivalent development using the Berthelot equation of state leads to a ρ_r/T dependence of δ, but in practice T is fairly constant over the temperature range of typically used for organic synthesis. (*72*)

The values of δ for the reactions in Table I at or below 60°C fall mostly in the range from 7-8. The low end of this scale is rougly equivalent to *n*-pentane (δ = 7.1) while the upper end corresponds to a somewhat weaker solvent (in the Hildebrand sense) than cyclohexane (δ = 8.2) and considerably weaker than CCl_4 (δ = 8.6). (*73*) Reactions in the 80-100°C range have even lower δ values at useful pressures, lower in fact than those of conventional organic solvents.

We can now define a practical pressure-temperature envelope for reactions in neat supercritical carbon dioxide using the curves in Figure 1. First, we note that operating pressures up to 6000 psi represent the limit for most systems built around HPLC pumps as well as the pressure limit for most commercial autoclaves. Pressures up to 10,000 psi are readily produced by some commercially available high pressure

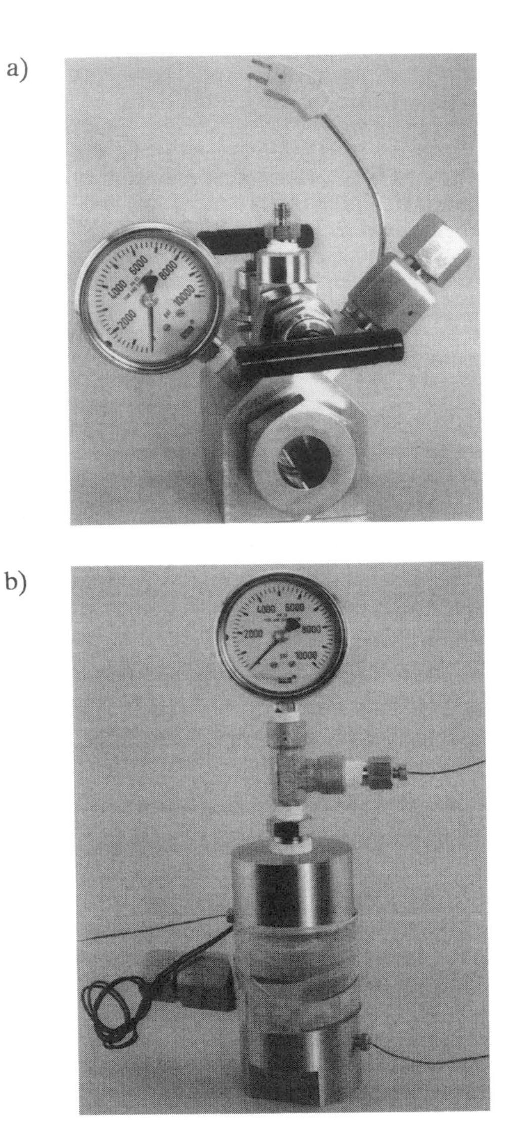

Figure 2: a) (above) Reaction cell with sapphire window used for pressures up to 6000 psi. b) (below) Windowless cell used for higher pressures.

syringe pumps and other devices. Moreover, in general, stirred reactors of a liter or more are difficult to obtain commercially with pressure ratings above 6000 psi and virtually impossible above 10,000 psi.

The discussion above suggests an approximate minimum value of $\delta = 7$ for most reactions. For neat supercritical CO_2, δ falls below 7 at a temperature of 83°C at 6000 psi and at 130°C at 10,000 psi using equation 6. The lower temperature limit is set by the critical temperature of CO_2: 31.1°C. Thus, for most workers, the useful range for synthesis in supercritical CO_2 is likely to be $31 < T < 83$°C, $P < 6000$ psi. The outer limit, without the use of fairly specialized and expensive equipment is $31 < T < 130$°C, $P < 10,000$ psi. If higher pressures are required, they may be most easily achieved using flow reactors, (which also scale more easily). Flow reactor methods for synthesis in supercritical carbon dioxide have been pioneered by Poliakoff and coworkers. (*74*) In practice, such reactors closely resemble those already used for hydrothermal processing.

Equation 6 indicates that the solvent strength, δ, is pressure-dependent, providing a potential route to improved selectivity and rate by "pressure-tuning" the solvent. A number of attempts to demonstrate reactivity control in supercritical CO_2 for Diels-Alder (*75-77*) and organic photoreactions (*78,79*) have exhibited very small effects. Andrew and coworkers have recently demonstrated dramatic solvent cage effects on selectivity of a photo-Fries reaction close to the critical density.(*80*) More polar SCF's have shown more promising results; control of esterification rates and polyester molecular weight distribution via enzymatic catalysis in fluoroform has been demonstrated. (*81,82*)

Experimental Considerations for Synthetic Chemistry in Supercritical CO_2

Below is a brief description of the equipment and methodology we have employed at Los Alamos to explore reactions in supercritical CO_2 and other SCF's.

1) Batch Reactors. We have performed most of the chemistry described here in 25-50 ml reaction cells which serve as "high pressure beakers." An important feature of these cells (shown in Figure 2a) is a window made of unoriented sapphire (Meller Optics, 1.25" diameter by 1 cm thick). The window allows the reaction to be monitored visually, and enables us to confirm that the reaction mixture is single-phase— a necessity for reactions near T_c. The windows are held in place by glands and sealed with 1/16" thick Teflon or 90 durometer Buna-N gaskets on both sides. (Viton and ethylene-propylene rubbers swell unacceptably when in contact with supercritical CO_2; Teflon is an acceptable substitute for Buna-N, paticularly if strong oxidizers are being used.)

These reaction cells can be fabricated inexpensively and customized for the gauge and valve connections required by a particular experiment. We routinely use the cells at pressures up to 6000 psi and temperatures up to 80°C, which covers the most useful range for reactions in supercritical CO_2, as discussed above. Similar reactors have been described by Lemert and DeSimone for use in polymerization reactions. (*83*)

At higher pressures, we have been compelled to use cells without windows such as those shown in Figure 2b. These cells operate up to 10,000 psi and 130°C, thereby covering what we view as the full range of useful reaction conditions in supercritical carbon dioxide. Small diamond windows which operate to even higher pressure and temperatures have been described in the literature for Raman spectroscopy (*84,85*) and transient grating experiments, (*86*) as well as for visual and optical monitoring phase behavior in supercritical water. (*87*) Gold gaskets are required to seal the windows.

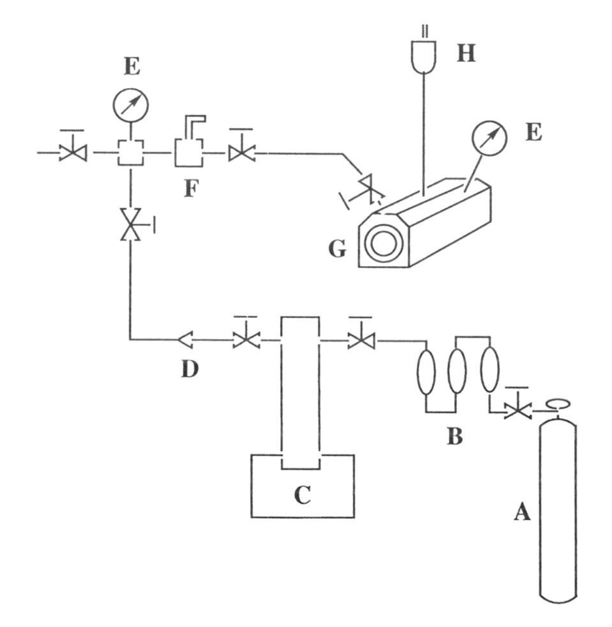

Figure 3: A schematic of a typical high pressure apparatus. **A:** high purity CO_2 tank; **B:** charcoal, oxysorb catalyst, drying columns; **C:** ISCO pump; **D:** check valve; **E:** pressure gauge; **F:** relief valve; **G:** sapphire window; **H:** thermocouple. Other valves on high pressure cell are omitted for clarity.

Both types of cells are small enough to fit on a small stirplate and agitation is achieved with a magnetic stir bar, avoiding the maintenance problems associated with mechanical stirrers in pressure vessels. The cells are heated with flexible heaters or heating tape connected to a temperature controller. Pressure relief is provided with either a relief valve or (preferably) a rupture disk.

An advantage of employing small reaction cells is that they can be easily connected to a high pressure manifold linked to a high pressure syringe pump. One pump can then supply pressurized CO_2 to a number of cells. A schematic of a high pressure manifold is shown in Figure 3. High purity CO_2 from tank **A** is introduced into pump **C** via columns **B** which remove organics, further dry, and deoxygenate the CO_2. The pump can then compress the CO_2 through check valve **D** into a manifold equipped with a pressure gauge **E** and relief valve **F**. The CO_2 is introduced into a view cell equipped with a sapphire window **G**, pressure gauge **E** and thermocouple **H** (other valves omitted for clarity). Heating tape is wrapped around the cell and is connected, along with the thermocouple, to a temperature controller. The cell sits on a magnetic stirplate and the solution inside is agitated with a stir bar. For safety, the cell is anchored in place and positioned behind protective shielding (0.5 inch polycarbonate). Air-sensitive catalysts or substrates can be added in sealed glass ampoules which break upon pressurization. (Tanko, J.M., Virginia Polytechnic Institute, personal communication, 1993).

2) Product Recovery. An important consideration in running reactions in supercritical CO_2 is letting down the reactor pressure in such a way as to ensure quantitative recovery of the reaction components. This problem has been addressed primarily by workers in the field of supercritical fluid extraction (SFE). (*88-93*) In general, the pressure is released through a long narrow tube (flow restrictor) whose tip is immersed in a collection solvent although solid-phase traps can be used. (*94*) The principal complication is that the solvent density falls as it moves down the restrictor (along with solute solubility) leading to potential problems of plugging or nonquantitative recovery. Joule-Thomson cooling during the depressurization can also decrease the solubility of the analyte and cause plugging. The latter effect can be counteracted by heating the restrictor, preferably as close to the tip as possible, which requires that the tip be placed just below the surface of the collection solvent.

A number of restrictor designs have been described in the literature, the most common type being linear restrictors. A linear restrictor is either a long capillary often used in chromatography or a 1/16" stainless steel tubing. (*88*) Linear restrictors are popular since they are inexpensive and can be heated to counteract the cooling effects during depressurization. The primary disadvantage in this system is that the tip of the tube cannot be heated during sample collection and plugging can occur.

Identification of Promising Reactions

Ideally, one would like to investigate reactions which will function as well or better in supercritical CO_2 as in conventional media. Unfortunately, reactions in supercritical CO_2 are not yet well enough understood to enable one to predict in advance which reactions can be translated most successfully into supercritical CO_2. However, considerations of solubility are clearly of primary importance. In order to achieve high concentrations, reagents should be neutral and relatively nonpolar . Similar solubility constraints affect the catalyst although these are less stringent and ionic catalysts can be employed. Reactions that can be carried out below 80°C are preferable. The most promising reactions would be those which capitalize on the properties of supercritical fluids and therefore offer the best opportunities for improved chemistry. In our view, the most promising reactions are those which, 1) work best in nonpolar solvents such as hexane, 2) are solvent-sensitive (and thus likely to be pressure-tunable), 3) lead to high-value products (mitigating the capital investment barriers to implementation of new technology), or 4) utilize gases which are not highly soluble in conventional solvents such as hydrogen or oxygen.

In addition, as discussed above, oxidation reactions and reactions which use CO_2 as a reagent as well as a solvent are worth investigating. Examples of both are discussed below. Finally, electrophilic processes may be advantageously transferred to supercritical CO_2, as demonstrated by the improved isomerization of C_4-C_{12} paraffins catalyzed by aluminum bromide. (*2,44*) Below, we describe three catalytic reactions which appear promising by these criteria: asymmetric catalytic hydrogenation of enamides, ruthenium-catalyzed two-phase oxidation of cyclohexene, and the catalytic copolymerization of CO_2 with epoxides.

Asymmetric Catalytic Hydrogenation of Enamides

Asymmetric catalytic hydrogenation, in which chiral catalysts are used to transform inexpensive achiral substrates into valuable chiral products via addition of H_2, is generally highly solvent-sensitive, owing to the small energy difference between the two diasteriomeric transition states. (*95-99*) Thus, this class of reactions meets the latter three of the four criteria above. In order to meet the first criterion, we focused on enantioselective hydrogenation of prochiral α-enamides **3** using cationic rhodium catalysts **1** which incorporate the chiral bidentate (*R*,*R*)-1,2-bis(*trans*-2,5-diethylphospholano)benzene ((*R*,*R*)-Et-DuPHOS) ligand **2**. (*100*) These cationic DuPHOS-Rh complexes have been found to catalyze the efficient hydrogenation of α-enamide esters **3** to valuable α-amino acid derivatives **4** with very high enantioselectivities (≥ 98% ee) in organic solvents, particularly hexane (equation 7). (*101,102*)

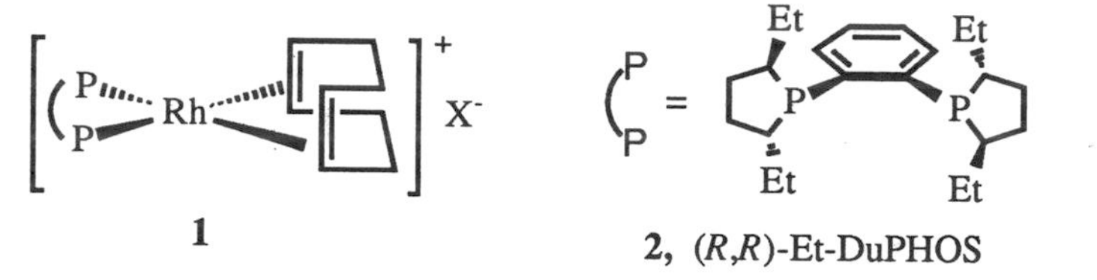

1

2, (*R*,*R*)-Et-DuPHOS

$$\text{R-CH=C(CO}_2\text{Me)N(H)Ac}\ (\mathbf{3}) \xrightarrow[H_2]{\mathbf{1}} \text{R-CH}_2\text{-CH(CO}_2\text{Me)N(H)Ac}\ (\mathbf{4}) \qquad (7)$$

Although the catalyst is cationic, we have found that it is soluble in both hexane and supercritical CO_2 if the highly lipophilic anion tetrakis(3,5-bis(trifluoromethyl)-phenyl)borate (*103*) ("BARF") is used as the counterion. Use of the trifluoromethanesulfonate counterion ($CF_3SO_3^-$), also provided sufficient solubility to allow catalysis to proceed in supercritical CO_2. The use of these anions may represent a general method for introducing cationic catalysts into supercritical CO_2, widening the potential range of reactions in which supercritical CO_2 can serve as a replacement solvent.

Hydrogenation reactions were performed in supercritical CO_2 by charging a 35 mL reactor of the type shown in Figure 2a with catalyst and enamide **3** (substrate/catalyst = 500), followed by pressurization with hydrogen gas (200 psig) and CO_2 (3000 psig). Addition of hydrogen affects the phase behavior of SC CO_2,

(*104*) but the medium was single-phase under the reaction conditions employed. A homogeneous supercritical phase (observed through the sapphire window) was produced (5000 psig total pressure) upon warming to 40°C , and the reactions were allowed to proceed for 24 hours. Product analysis by ^1H and ^{13}C NMR, as well as mass spectrometry, was used to confirm complete hydrogenation. We have also conducted each hydrogenation reaction in methanol and hexane at the same temperature and reaction time to compare our results in supercritical CO_2. Hexane was selected to serve as a condensed-phase model for the nonpolar environment of supercritical CO_2.

Table II lists the enantioselectivities achieved in the hydrogenation of four α-enamides substrates **3a-d** using the catalyst **1** in hexane, methanol and supercritical CO_2. (Reaction conditions in methanol or *n*-hexane: S/C = 500, 60 psig H_2, 40°C, 24 h. Reaction conditions in supercritical CO_2: S/C = 500, 200 psig H_2, 5000 psig (overall), 40°C, 24 h.) The reduction of enamides **3** proceeded cleanly and quantitatively to provide the α-amino acid derivatives (*R*)-**4** with high enantioselectivity (90.9-99.7% ee) in each solvent. As shown in Table II, the enantioselectivities achieved with this catalyst system in supercritical CO_2 are very high, and are comparable to those achieved in conventional solvents. (*101,102*)

Table II: Enantioselectivities for Reaction 8 Using Catalyst 1

R	X^-	solvent		
		MeOH	Hexane	SC CO_2
H	BARF	98.7	96.2	99.5
H	$CF_3SO_3^-$	99.4	99.1	99.1
Et	BARF	98.7	96.8	98.8
Et	$CF_3SO_3^-$	99.7	99.6	98.8
Ph	BARF	97.5	98.3	99.1
Ph	$CF_3SO_3^-$	99.0	98.7	90.9
3,5-$(CF_3)_2C_6H_3$	BARF	93.2	96.6	91.9
3,5-$(CF_3)_2C_6H_3$	$CF_3SO_3^-$	99.1	98.6	94.6

SOURCE: Adapted from ref. 38.

Another class of enamides, β,β-disubstituted α-enamide esters, have proven difficult to reduce with high enantioselectivity. For example, the highest enantioselectivity reported for the hydrogenation of dehydrovaline derivative **5** in conventional solvents was 55% ee using a cationic DIPAMP-Rh catalyst. (*105*) We have found that the Et-DuPHOS-Rh catalysts **1** outperform all other reported catalysts in the hydrogenation of two representative β,β-disubstituted α-enamide esters, **5** and **7** (equations 8 and 9). Hydrogenation of enamide **5** in MeOH (60 psig H_2, 40°C) led to the valine derivative (*R*)-**6**, in 62.6% and 67.4% ee using BARF and $CF_3SO_3^-$ counterions respectively. Hydrogenation of **5** in supercritical CO_2 (40°C, 5000 psi, 24 hours, S/C = 500) gave significantly better enantioselectivities than either hexane or methanol: 84.7% for the BARF salt and 88.4% for $CF_3SO_3^-$. Equally dramatic gains in enantioselectivity were observed in the hydrogenation of another β,β-disubstituted α-enamide ester, **7**, (see Table III.)

Although the unique solvating properties of supercritical CO_2 may be the cause of the enantioselectivity enhancements observed, differences in selectivity can be caused by pressure in conventional solvents as well. For this reason, we carried out two hydrogenation reactions of **5** in hexane at 40°C with catalyst **1** as the BARF

$$\text{5 (N(H)Ac, } CO_2Me\text{)} \xrightarrow[\text{1, 40°C, 24 h.}]{H_2/SC\ CO_2} \text{6 (N(H)Ac, } CO_2Me\text{)} \quad (8)$$

$$\text{7 (N(H)Ac, } CO_2Me\text{)} \xrightarrow[\text{1, 40°C, 24 h.}]{H_2/SC\ CO_2} \text{8 (N(H)Ac, } CO_2Me\text{)} \quad (9)$$

Table III. Enantioselectivities from Hydrogenation of 5 and 7 using 1

Substrate	X^-	solvent		
		MeOH	Hexane	SC CO_2
5	BARF	62.6	69.5	84.7
5	$CF_3SO_3^-$	67.4	70.4	88.4
7	BARF	81.1	76.2	96.8
7	$CF_3SO_3^-$	95.0	91.2	92.5

SOURCE: Adapted from ref. 38.

salt, one under 200 psi H_2 pressure, and the other under 200 psi H_2 and a total pressure of 5000 psi with nitrogen gas. Enantioselectivities observed under these conditions were 69.5% ee and 74.5% ee, respectively. These results indicate that simple pressure effects are not responsible for high ee's in the hydrogenation of **5** with catalysts **1**, and suggest that selectivity enhancement is specifically associated with the use of supercritical CO_2 as a reaction solvent.

Phase Transfer Catalytic Oxidation in an Aqueous/Supercritical CO_2 Medium

Supercritical carbon dioxide represents a potential replacement for organic solvents used, along with water, as media for phase transfer catalytic oxidation. The process involves the use of an aqueous phase, where the terminal oxidant resides in contact with an organic phase, consisting either of the neat substrate or a solution of the substrate. The catalyst shuttles between the two phases. This approach to oxidation eliminates the need for highly polar solvents (which are typically expensive and hard to separate) used in one-phase catalytic methods while requiring only catalytic amounts of precious metals such as osmium or ruthenium. (*106,107*) Such syntheses typically use halogenated solvents in order to minimize the organic solvent's solubility in the aqueous phase. CO_2 is highly immiscible with water even at high pressure, making it an attractive substitute solvent. (*108,109*) An unavoidable limitation of reactions in CO_2/water systems, however, is that the aqueous phase is forced to roughly pH 5 by the equilibrium between water, CO_2 and carbonic acid (Equation 10). (*110*) Consequently, base-catalyzed or other two-phase reactions which require basic conditions are not candidates for solvent substitution by supercritical CO_2.

$$CO_2 + H_2O \rightleftharpoons H_2CO_3 \rightleftharpoons H^+ + HCO_3^- \quad (10)$$

We have attempted to translate the two-phase oxidation of alcohols, alkenes, and other substrates mediated by high valent ruthenium oxo complexes from water/halocarbon media to water/supercritical CO_2. In this system, $RuCl_3$, dissolved in the aqueous phase, is oxidized to RuO_4 which transfers to the organic phase where it reacts with the substrate. The volatility and lack of polarity of RuO_4 suggest that it should have acceptable solubility in supercritical CO_2. RuO_4 is such a strong oxidant that it cannot be used without a solvent, (*111*) generally carbon tetrachloride. Many oxidizing agents have been used, including NaOCl, (*112*) $NaBrO_3$, (*113*) and $NaIO_4$. (*114*) In particular, oxidation of cyclic alkenes (cyclopentene (*115*) or cyclohexene (*111*)) gives the corresponding diacids almost exclusively. Because of its economic importance, we chose to focus on the ruthenium-mediated oxidation of cyclohexene to adipic acid (equation 11).

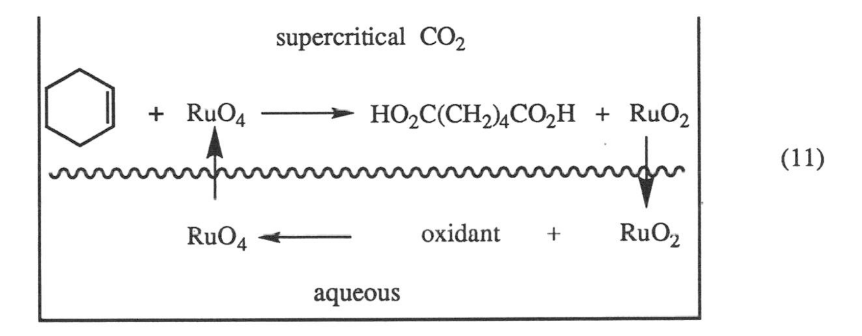

(11)

Reactions were performed in a 30 ml viewcell similar to that shown in Figure 2a using 4 mmol of cyclohexene, 17 mmol of $NaIO_4$ and 0.1 mmol of catalyst (2.5% vs. substrate). Reactions were conducted at 40°C and 2400 or 4000 psi for 12-24 hours. Selectivity to adipic acid was almost always 99% or better, but in no case did the catalyst turn over more than five times. Use of other oxidants including Ce(IV) and peroxyacetic acid gave similar results. When NaOCl was employed as the oxidant led primarily to 1,2-dichlorocyclohexane, owing to the acidity of the medium.

The limited number of turnovers in oxidations leading to carboxylic acids is a common occurence in ruthenium-catalyzed oxidations and is usually ascribed to complexation of a Ru(IV) intermediate by the acid. Addition of acetonitrile to the aqueous phase normally disrupts the complex and restores catalytic activity. (*116,117*) In this case, however, addition of acetonitrile did not improve turnover number. We tentatively ascribe the slow rates to interference by bicarbonate which may complex Ru(IV) strongly enough to prevent its liberation by acetonitrile. Although the retention of high selectivity is encouraging, this work suggests that translating two-phase reactions from organic/H_2O to supercritical CO_2/H_2O can be frustrated by species arising from the equilibrium in equation 10.

Catalytic Copolymerization of CO_2 With Epoxides

Supercritical CO_2 may prove to be a particularly advantageous reaction medium when CO_2 serves as both a reagent and a solvent. (*16*) A novel synthesis of ethylene complexes in supercritical ethylene (*118*) and the improved rates for catalytic hydrogenation of CO_2 to formic acid discussed above (equation 1) both provide support for this approach. (*35-37*) Another class of reactions which utilizes CO_2 is the catalytic copolymerization of epoxides with CO_2 to form polycarbonates (equation 12). (*119-131*) Such reactions have been investigated previously in supercritical

CO_2. (*132*) The principal complications of this reaction are very low rates and the formation of substantial amounts of cyclic carbonates as byproducts. An interesting class of catalysts based on adducts of di- and triphenols to diethylzinc, developed by Kuran, (*133,134*) is reported to have a rate and selectivity which increase with CO_2 concentration, suggesting that both can be improved by operating in supercritical CO_2 instead of the usual solvent, dioxane. The highest selectivity in dioxane is reported for the 3:2 diethylzinc: catechol adduct.

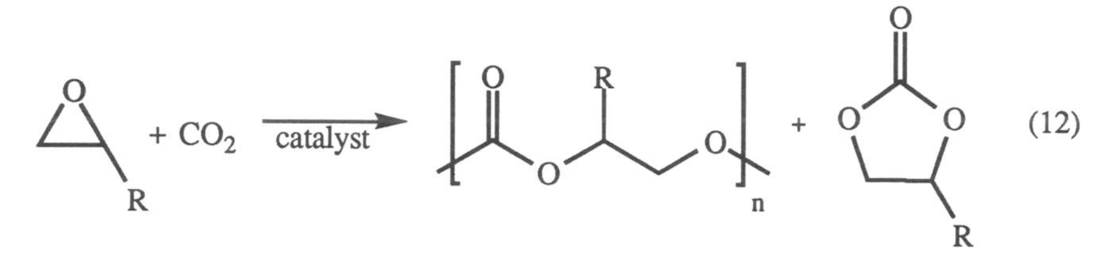

Reactions were performed in 300 ml stirred autoclaves at 5000 psi, 50°C for 24 hours using 50-70 ml of dry toluene, THF or dioxane as a cosolvent. Typically, 0.6 mmol of Zn as the 3:2 Et_2Zn/catechol adduct was used along with 70 mmol of epoxide (epoxide:Zn = 117. Lit. = 25:1 in dioxane). Conversions to polymer were never more than 1%, indicating no substantial increase in rate above that achievable in conventional solvents under a few atmospheres of CO_2. Products were isolated by the same procedures used for literature reactions in conventional solvents.

IR analysis indicated no substantial improvement in polymer to cyclic carbonate selectivity. The best result achieved was 2.3 in the case of 1,2-butylene oxide compared to 2.0 for propylene oxide copolymerization in dioxane (the only literature result for this catalyst). (*134*) The polymer produced (in the case of cyclohexene oxide) was of moderate molecular weight and highly polydisperse (M_n = 13,300, M_w = 68,800, M_w/M_n = 5.2 using polystyrene standards).

The low rates observed in supercritical CO_2 are most likely due to the fact that CO_2 is a much weaker solvent than dioxane. Despite the failure to observe enhanced reactivity in this system, the success of Noyori and coworkers suggests that investigation of other CO_2 reactions, particularly polymerizations, (*135,136*) in supercritical CO_2 is worthwhile. Many such reactions remain unexplored in supercritical CO_2 including catalytic reactions between butadiene and CO_2, (*137*) formation of isocyanates, (*138,139*) urethanes, (*139-143*) ureas, (*144,145*) and polyureas (*146,147*) from CO_2 and amines, and production of cyclic carbonates from CO_2 and epoxides.

Conclusions

Supercritical CO_2 can be a viable substitute solvent for many of synthetic reactions. The use of ionic catalysts at low concentration is possible if a suitable counterion is chosen; BARF has been especially successful. The apparatus described in this chapter is can be easily fabricated allowing for investigation of a wide range of reactions in supercritical solvents.

In our view, the best candidates for useful reactions in CO_2 are those that: 1) work best in nonpolar solvents and can be run at temperatures below 80°c, 2) are solvent-sensitive and thus likely to be "pressure-tuneable," 3) lead to high-value products (mitigating the capital investment barriers to implementation of new technology), or 4) utilize gaseous reagents which are not highly soluble in conventional solvents. There is significant opportunity for improved rates and selectivity as demonstrated by results in hydrogenation by Noyori (*35-37*) and from our laboratory. Another promising class of reagents are those which use CO_2 as both a solvent and reagent. Implementation of two-phase (aqueous/organic) reactions must take into account the acidity of the aqueous medium due to the carbonic acid equilibrium.

Acceptance of supercritical fluids for solvent replacement in the chemical industry will probably require economic as well as environmental drivers. A key component of research in supercritical fluids must therefore be to identify and quantify the advantages (e.g. faster rates, higher selectivities) of carrying out chemical transformations in supercritical CO_2 rather than in conventional solvents. The development of other reactions in supercritical CO_2 with superior rates and selectivities sufficient to justify the costs of implementing SCF-based production processes is essential to realizing pollution prevention by this route.

Acknowledgments

We thank Dr. Paul Anastas, Dr. Carol Burns, Dr. Steve Buelow, and Dr. Dale Spall for helpful discussions. Part of this work was supported by Department of Energy LDRD funding through Los Alamos National Laboratory and another part by the US EPA Office of Pollution Prevention and Toxics' Design for the Environment Program.

Literature Cited

1. Wu, B. C.; Paspek, S. C.; Klein, M. T.; LaMarca, C. In *Supercritical Fluid Technology: Reviews in Modern Theory and Applications*; Bruno, T. J. and Ely, J. F., Ed.; CRC Press: Boca Raton, 1991, pp 511.
2. Caralp, M. H. M.; Clifford, A. A.; Coleby, S. E. In *Extraction of Natural Products Using Near-Critical Solvents*; King, M. B. and Bott, T. R., Ed.; Blackie: Glasgow, U.K., 1993, pp 50.
3. Clifford, A. A. In *Supercritical Fluids, Fundamentals for Application*; Kiran, E. and Levelt Sengers, J. M. H., Ed.; Kluwer: Dordrecht, 1994, pp 449.
4. Kaupp, G. *Angew. Chem. Int. Ed. Engl.* **1994**, *33*, 1452.
5. Jessop, P. G.; Ikariya, T.; Noyori, R. *Science* **1995**, *269*, 1065.
6. Johnston, K. P. In *Supercritical Fluid Science and Technology*; Johnston, K. P. and Penninger, J., Ed.; ACS Symposium Series; American Chemical Society: Washington DC, 1989; Vol. 406, pp 1.
7. Wu, B. C.; Klein, M. T.; Sandler, S. I. *Ind. Eng. Chem. Res.* **1991**, *30*, 822.
8. Brennecke, J. F. In *Supercritical Fluid Engineering Science*; Kiran, E. and Brennecke, J. F., Ed.; ACS Symposium Series; American Chemical Society: Washington D.C., 1993; Vol. 514, pp 201.
9. Vesovic, V.; Wakeham, W. A.; Olchowy, G. A.; Sengers, J. V.; Watson, J. T. R.; Millat, J. *J. Phys. Chem. Ref. Data* **1990**, *19*, 763.
10. Sovova, H.; Prochazka, J. *Ind. Eng. Chem. Res.* **1993**, *32*, 3162.
11. Debenedetti, P. G.; Reid, R. C. *AIChE J.* **1986**, *32*, 2034.
12. Sassiat, P. R.; Mourier, P.; Caude, M. H.; Rosset, R. H. *Anal. Chem.* **1987**, *59*, 1164.
13. Liong, K. K.; Wells, P. A.; Foster, N. R. *Ind. Eng. Chem. Res.* **1992**, *31*, 390.
14. Levelt-Sengers, J. M. H.; Deiters, U. K.; Klask, U.; Swidersky, P.; Schneider, G. M. *Int. J. Thermophys.* **1993**, *14*, 893.
15. Catchpole, O. J.; King, M. B. *Ind Eng. Chem. Res.* **1994**, *33*, 1828.
16. Reetz, M. T.; Konen, W.; Strack, T. *Chimia* **1993**, *47*, 493.

17. Levelt Sengers, J. M. H. In *Supercritical Fluid Technology, Reviews in Modern Theory and Applications*; Bruno, T. J. and Ely, J. F., Ed.; CRC Press: Boca Raton, 1991, pp 1.
18. Eckert, C. A.; Ziger, D. H.; Johnston, K. P.; Kim, S. *J. Phys. Chem.* **1986**, *90*, 2738.
19. Yonker, C. R.; Frye, S. L.; Kalkwarf, D. R.; Smith, R. D. *J. Phys. Chem.* **1986**, *90*, 3022.
20. Mori, Y.; Shimizu, T.; Iwai, Y.; Arai, Y. *J. Chem. Eng. Ref. Data* **1992**, *37*, 317.
21. Kim, S.; Johnston, K. P. *Ind. Eng. Chem. Res.* **1987**, *26*, 1206.
22. Brennecke, J. F.; Tomasko, D. L.; Peshkin, J.; Eckert, C. A. *Ind. Eng. Chem. Res.* **1990**, *29*, 1682.
23. Novotny, M.; Bertsch, W.; Zlatkis, A. *J. Chromatogr.* **1971**, *61*, 17.
24. Klesper, E. *Angew. Chem. Int. Ed. Engl.* **1978**, *17*, 738.
25. van Wasen, U.; Swaid, I.; Schneider, G. M. *Angew. Chem. Int. Ed. Engl.* **1980**, *19*, 575.
26. Fields, S. M.; Lee, M. L. *J. Chromatogr.* **1985**, *349*, 305.
27. Lee, M. L.; Markides, K. E. *Science* **1987**, *235*, 1342.
28. Dean, J. R. ed. *Application of Supercritical Fluids in Industrial Analysis*; CRC Press: Boca Raton, 1994.
29. Bok, E.; Kelch, D.; Schumacher, K. S. *Solid State Tech.* **1992**, *35*, 117.
30. Novak, R. A.; Reightler, W. J.; Robey, R. J.; Wildasin, R. E. *Int. J. Environment. Conscious Design Manufact.* **1993**, *2*, 29.
31. Spall, W. D. *Int. J. Environment. Conscious Design Manufact.* **1993**, *2*, 81.
32. Helling, R. K.; Tester, J. W. *Env. Sci and Tech.* **1988**, *22*, 1319.
33. Stadig, W. P. *Chem. Processing* **1995**, *August*, 34.
34. DeSimone, J. M.; Guan, Z.; Elsbernd, C. S. *Science* **1992**, *257*, 945.
35. Jessop, P. G.; Ikariya, T.; Noyori, R. *Nature* **1994**, *368*, 231.
36. Jessop, P. G.; Hsiao, Y.; T., I.; Noyori, R. *J. Am. Chem. Soc.* **1994**, *116*, 8851.
37. Jessop, P. G.; Hsiao, Y.; Ikariya, T.; Noyori, R. *J. Chem. Soc. Chem. Commun.* **1995**, 707.
38. Burk, M. J.; Feng, S.; Gross, M. F.; Tumas, W. *J. Am. Chem. Soc.* **1995**, 8277.
39. Terry, R. E.; Zaid, A.; Angelos, C.; Whitman, D. L. *Energy Prog.* **1988**, *8*, 48.
40. DeSimone, Y. M.; Maury, E. E.; Lemert, R. E.; Combes, J. R. *Polym. Mater. Sci. Eng.* **1993**, *68*, 41.
41. Guan, Z.; Combes, J. R.; Menceloglu, Y. Z.; DeSimone, J. M. *Macromolecules* **1993**, *26*, 2663.
42. DeSimone, J. M.; Maury, E. E.; Menceloglu, Y. Z.; McClain, J. B.; Romack, T. J.; Combes, J. R. *Science* **1994**, *265*, 356.
43. Adamsky, F. A.; Beckman, E. J. *Macromolecules* **1994**, *27*, 312.
44. Kramer, G. M.; Leder, F. (Exxon), US Patent 3,889,945, 1975.
45. Leder, F.; Kramer, G. M.; Solomon, H. J. (Exxon), US Patent 3,946,088, 1976.
46. Randolph, T. W.; Blanck, H. W.; Prausnitz, J. M. *AIChE J.* **1988**, *34*, 1354.
47. Randolph, T. W.; Clark, D. S.; Blanch, H. W.; Prausnitz, J. M. *Proc. Natl. Acad. Sci. USA* **1988**, *85*, 2979.
48. Aaltonen, O.; Rantakyla, M. *Chemtech* **1991**, *21*, 240.
49. Cenia, E.; Palocci, C.; Gasparrini, F.; Misiti, D.; Fagnano, N. *J. Mol. Catal.* **1994**, *89*, L11.
50. Jobling, M.; Howdle, S. M.; Healy, M. A.; Poliakoff, M. *J. Chem. Soc. Chem. Commun.* **1990**, 1287.

51. Poliakoff, M.; Howdle, S. M.; Kazarian, S. G. *Angew. Chem. Int. Ed. Engl.* **1995**, *34*, 2001.
52. Tanko, J. M.; Blackert, J. F. *Science* **1994**, *263*, 203.
53. Rathke, J. W.; Klingler, R. J.; Krause, T. B. *Organometallics* **1991**, *10*, 1350.
54. Bartle, K. D.; Baulch, D. L.; Clifford, A. A.; Coleby, S. E. *J. Chromatogr.* **1991**, *557*, 69.
55. Chrastil, J. *J. Phys. Chem.* **1982**, *86*, 3016.
56. Janssen, H.-G.; Cramers, C. A. *J. Chromatogr.* **1990**, *505*, 19.
57. Ekart, M. P.; Bennett, K. L.; Eckert, C. A. In *Supercritical Fluid Engineering Science: Fundamentals and Applications*; Kiran, E. and Brennecke, J. F., Ed.; ACS Symposium Series; American Chemical Society: Washington D.C., 1993; Vol. 514, pp 228.
58. Phillips, J. H.; Robey, R. J. *J. Chromatogr.* **1989**, *465*, 177.
59. Newman, D. A.; Hoefling, T. A.; Beitle, R. R.; Beckman, E. J. *J. Supercrit. Fluids* **1993**, *6*, 165 and 205.
60. Hoefling, T. A.; Beitle, R. R.; Enick, R. M.; Beckman, E. J. *Fluid Phase Equilibria* **1993**, *83*, 203.
61. Lin, Y.; Brauer, R. D.; Laintz, K. E.; Wai, C. M. *Anal. Chem.* **1993**, *65*, 2549.
62. Laintz, K. E.; Wai, C. M.; Yonker, C. R.; Smith, R. D. *Anal. Chem.* **1992**, *64*, 2875.
63. Boatwright, D. L.; Suleiman, D.; Dillow-Wilson, A.; Eckert, C. A.; Liotta, C. L. *Ind. Eng. Chem. Res.* in press.
64. Bieling, V.; Kurz, F.; Rumpf, B.; Maurer, G. *Ind. Eng. Chem. Res.* **1995**, *34*, 1449.
65. Francis, A. W. *J. Phys. Chem.* **1954**, *58*, 1099.
66. Dandge, D. K.; Heller, J. P.; Wilson, K. V. *Ind. Eng. Chem. Prod. Res. Dev.* **1985**, *24*, 162.
67. Oostdyk, T. S.; Grob, R. L.; Snyder, J. L.; McNally, M. E. *J. Chromatogr. Sci.* **1993**, *31*, 177.
68. Fields, K. M.; Grolimund, K. *J. High Res. Chromatogr. and Chromatogr. Comm.* **1988**, *11*, 727.
69. Pitzer, K. S. *J. Am. Chem. Soc.* **1955**, *77*, 3427.
70. Pitzer, K. S.; Lippmann, D. Z.; Curl, R. F. J.; Huggins, C. M.; Petersen, D. E. *J. Am. Chem. Soc.* **1955**, *77*, 3433.
71. Drews, M. J.; Ivey, K.; Lam, C. *Text. Chem. Color.* **1994**, *26*, 29.
72. Giddings, J. C.; Myers, M. N.; McLaren, L.; Keller, R. A. *Science* **1968**, *162*, 67.
73. Barton, A. F. M. *CRC Handbook of Solubility Parameters and Other Cohesion Parameters*; 2nd ed.; CRC Press: Boca Raton, 1994.
74. Banister, J. A.; Lee, P. D.; Poliakoff, M. *Organometallics* in press.
75. Kim, S.; Johnston, K. P. *Chem. Eng. Commun.* **1988**, *63*, 49.
76. Ikushima, Y.; Saito, N.; Arai, M. *Bull. Chem. Soc. Japan* **1991**, *64*, 282.
77. Ikushima, Y.; Saito, N.; Sato, O.; Arai, M. *Bull. Chem. Soc. Japan* **1994**, *67*, 1734.
78. Hrnjez, B. J.; Mehta, A. J.; Fox, M. A.; Johnston, K. P. *J. Am. Chem. Soc.* **1989**, *111*, 2662.
79. Morgenstern, D. A.; Tumas, W. *J. Am. Chem. Soc.* submitted.
80. Andrew, D.; Des Islet, B. T.; Margaritis, A.; Weedon, A. C. *J. Am. Chem. Soc.* **1995**, *117*, 6132.
81. Kamat, S.; Iwascewyz, B.; Beckman, E. J.; Russell, A. J. *Proc. Natl. Acad. Sci. USA* **1993**, *90*, 2940.
82. Chaudhary, A. K.; Beckman, E. J.; Russell, A. J. *J. Am. Chem. Soc.* **1995**, *117*, 3728.

83. Lemert, D. M.; DeSimone, J. M. *J. Supercrit. Fluids* **1991**, *4*, 186.
84. Frantz, J. D.; Dubessy, J.; Mysen, B. *Chem. Geol.* **1993**, *106*, 9.
85. Foy, B. R.; Masten, D. A.; Harradine, D. M.; Dyer, R. B.; Buelow, S. J. *J. Phys. Chem.* **1993**, *97*, 8557.
86. Butenhoff, T. J. *Int. J. Thermophys.* **1995**, *16*, 1.
87. Anderson, G. K. ; Proceedings of the International Conference on the Properties of Water and Steam; Orlando.
88. Burford, M. D.; Hawthorne, S. B.; Miller, D. J.; Braggins, T. *J. Chromatogr.* **1992**, *609*, 321.
89. Wong, J. M.; Kado, N. Y.; Kuzmicky, P. A.; Ning, H.-S.; Woodrow, J. E.; Hsieh, D. P. H.; Seiber, J. N. *Anal. Chem.* **1991**, *63*, 1644.
90. Wright, B. W.; Wright, C. W.; Gale, R. W.; Smith, R. D. *Anal. Chem.* **1987**, *59*, 38.
91. Campbell, R. M.; Lee, M. L. *Anal. Chem.* **1986**, *58*, 2247.
92. Smith, R. D.; Fulton, J. L.; Petersen, R. C.; Kopriva, A. J.; Wright, B. W. *Anal. Chem.* **1986**, *58*, 2057.
93. "Application Bulletin No. 77," Isco, 1994.
94. Mulcahey, L. J.; Hedrick, J. L.; Taylor, L. T. *Anal. Chem.* **1991**, *63*, 2225.
95. Parshall, W.; Nugent, W. A. *Chemtech* **1988**, *18*, 184, 314, and 376.
96. Collins, A. N.; Sheldrake, G. N.; Crosby, J. *Chirality in Industry*; Wiley: New York, 1992.
97. Nugent, W. A.; RajanBabu, T. V.; Burk, M. J. *Science* **1993**, *259*, 479.
98. Ojima, I. *Catalytic Asymmetric Synthesis*; VCH: Weinheim, Germany, 1993.
99. Noyori, R. *Asymmetric Catalysis in Organic Synthesis*; Wiley: New York, 1994.
100. Burk, M. J.; Harper, G. P.; Kalberg, C. S. *J. Am. Chem. Soc.* **1995**, *117*, 4423.
101. Burk, M. J. *J. Am. Chem. Soc.* **1991**, *113*, 8518.
102. Burk, M. J.; J.E., F.; Nugent, W. A.; Harlow, R. L. *J. Am. Chem. Soc.* **1993**, *115*, 10125.
103. Brookhart, M.; Grant, B.; Volpe, A. F. *Organometallics* **1992**, *11*, 3920.
104. Tsang, C. Y.; Streett, W. B. *Chem. Eng. Sci.* **1981**, *36*, 993.
105. Scott, J. W.; Keith, D. D.; Nix, G.; Parrish, D. R.; Remington, S.; Roth, G. P.; Townsend, J. M.; Valentine, D. J.; Yang, R. *J. Org. Chem.* **1981**, *46*, 5086.
106. Courtney, J. L. In *Organic Syntheses by Oxidation with Metal Compounds*; Mijs, W. J. and de Jonge, C. R. H. I., Ed.; Plenum: New York, 1986, pp 445.
107. Dehmlow, E. V.; Dehmlow, S. S. *Phase Transfer Catalysis*; 3rd ed.; VCH: Weinheim, 1993.
108. Crovetto, R.; Wood, R. H. *Fluid Phase Equilibria* **1992**, *74*, 271.
109. Gallagher, J. S.; Crovetto, R.; Levelt Sengers, J. M. H. *J. Phys. Chem. Ref. Data* **1993**, *22*, 431.
110. Kruse, R.; Franck, E. U. *Ber. Bunsenges. Phys. Chem.* **1982**, *86*, 1036.
111. Berkowitz, L. M.; Rylander, P. N. *J. Am. Chem. Soc.* **1958**, *80*, 6682.
112. Wolfe, S.; Hasan, S. K.; Campbell, J. R. *J. Chem. Soc.* **1970**, 1420.
113. Giddings, S.; Mills, A. *J. Org. Chem.* **1988**, *53*, 1103.
114. Spitzer, U. A.; Lee, D. G. *J. Org. Chem.* **1975**, *40*, 2539.
115. Orita, H.; Hayakawa, T.; Takehira, K. *Bull. Chem. Soc. Japan* **1986**, *59*, 2637.
116. Rossiter, B. E.; Katsuki, T.; Sharpless, K. B. *J. Am. Chem. Soc.* **1981**, *103*, 464.
117. Carlsen, P. H. J.; Katsuki, T.; Martin, V. S.; Sharpless, K. B. *J. Org. Chem.* **1981**, *46*, 3936.
118. Banister, J. A.; Howdle, S. M.; Poliakoff, M. *J. Chem. Soc. Chem. Commun.* **1993**, 1814.

119. Stevens, H. (Pittsburgh Plate Glass Co.), US Patent 3248415, 1966.
120. Inoue, S.; Koinuma, H.; Tsuruta, T. *Makromol. Chem.* **1969**, *130*, 210.
121. Inoue, S.; Koinuma, H.; Yokoo, Y.; Tsuruta, T. *Makromol. Chem.* **1971**, *143*, 97.
122. Kobayashi, M.; Inoue, S.; Tsuruta, T. *Macromolecules* **1971**, *4*, 658.
123. Kobayashi, M.; Inoue, S.; Tsuruta, T. *J. Polym. Sci.* **1973**, *11*, 2383.
124. Inoue, S. *Chemtech* **1976**, 588.
125. Soga, K.; Uehishi, K.; Hosoda, S.; Ikeda, S. *Makromol. Chem.* **1977**, *178*, 893.
126. Koinuma, H.; Hirai, H. *Makromol. Chem.* **1977**, *178*, 1283.
127. Takeda, N.; Inoue, S. *Makromol. Chem.* **1978**, *179*, 1377.
128. Soga, K.; Hyakkoku, K.; Ikeda, S. *Makromol. Chem.* **1978**, *179*, 2837.
129. Soga, K.; Uenishi, K.; Ikeda, S. *J. Polym. Sci.* **1979**, *17*, 415.
130. Inoue, S.; Takada, T.; Tatsu, H. *Makromol. Chem., Rapid Commun.* **1980**, *1*, 775.
131. Aida, T.; Inoue, S. *Macromolecules* **1982**, *15*, 682.
132. Petersen, U. *Methoden Org. Chem. (Houben Weyl)* **1983**, 95.
133. Kuran, W. *Makromol. Chem.* **1976**, *177*, 11.
134. Kuran, W. *Appl. Organomet. Chem.* **1991**, *5*, 191.
135. Rokicki, A.; Kuran, W. *J. Macromol. Sci.-Rev. Macromol. Chem.* **1981**, *C21*, 135.
136. Tsuda, T.; Hokazono, H. *Macromolecules* **1994**, *27*, 1289.
137. Hoberg, H.; Barhausen, D. *J. Organomet. Chem.* **1989**, *379*, C7.
138. Waldman, T. E.; McGhee, W. D. *J. Chem. Soc. Chem. Commun.* **1994**, 957.
139. Riley, D.; McGhee, W. D.; Waldman, T. In *Benign by Design: Alternative Synthetic Design for Pollution Prevention*; Anastas, P. T. and Farris, C. A., Ed.; ACS Symposium Series; American Chemical Society: Washington D.C., 1994; Vol. 577, pp 122.
140. Riley, D. P.; Christ, M. E.; Christ, K. M. *Organometallics* **1993**, *12*, 1429.
141. Aresta, M.; Quaranta, E. *Tetrahedron* **1992**, *48*, 1515.
142. McGhee, W. D.; Pan, Y.; Riley, D. P. *J. Chem. Soc. Chem. Commun.* **1994**, 699.
143. McGhee, W.; Riley, D.; Christ, K.; Pan, Y.; Parnas, B. *J. Org. Chem.* **1995**, *60*, 2820.
144. Ogura, H.; Takeda, K.; Tokue, R.; Kobayashi, T. *Synthesis* **1978**, 394.
145. Fournier, H.; Bruneau, C.; Dixneuf, P. H.; Lecolier, S. *J. Org. Chem.* **1991**, *56*, 4456.
146. Buckley, G.; Ray, N. (ICI), US Patent 2550767, 1951.
147. Rokicki, G. *Makromol. Chem.* **1988**, *189*, 2513.

RECEIVED November 7, 1995

Chapter 12

Reduction of Volatile Organic Compound Emissions During Spray Painting

A New Process Using Supercritical Carbon Dioxide to Replace Traditional Paint Solvents

M. D. Donohue[1], J. L. Geiger[1], A. A. Kiamos[1], and K. A. Nielsen[2]

[1]Department of Chemical Engineering, Johns Hopkins University, Baltimore, MD 21218
[2]Union Carbide Corporation, South Charleston, WV 25303

A new process for spraying paints and other coatings has been developed which reduces atmospheric emissions of environmentally harmful volatile organic compounds (VOCs). The liquid solvents of conventional coatings have been replaced by supercritical carbon dioxide. The carbon dioxide not only reduces viscosity, but provides additional benefits. The spray takes on a parabolic shape, is more finely atomized than conventional sprays, and has a narrower range of droplet sizes. The resulting coatings have uniform thickness and excellent coalescence. The thermodynamics, phase behavior, and rheology of carbon dioxide-solvent-polymer systems have been investigated. Data are presented that will help to reformulate conventional paints to be used with this process. While much remains to be learned about the thermodynamics and rheology of polymers mixed with supercritical fluids, it appears that this technology can be used with nearly all paints and coatings that currently are sprayed using organic solvents.

Volatile organic compounds (VOCs) are a class of air pollutants that enter the environment from many sources, including transportation, power generation, chemical manufacturing, and solvent evaporation during the spray application of coatings, paints, adhesives, and other materials. Every year, American industries spray 1.5 billion liters of coatings and paints, which release an average of 550 grams of VOC solvents for each liter sprayed (4.5 lb/gal). To put this into perspective, the VOC emissions from painting an automobile are greater than from the engine exhaust over its lifetime. As many VOC solvents are hazardous air pollutants, this represents a considerable burden on the environment and, consequently, much effort has been directed toward developing new coating systems to reduce VOC emissions.

0097–6156/96/0626–0152$12.00/0

A conventional solvent-borne coating has two basic components: materials that comprise the solid coating and organic solvents that allow for its application. The solid coating consists of one or more polymers that bind together and form a smooth film. Common coating polymers are acrylics, alkyds, polyesters, melamines, cellulosics, silicones, vinyls, epoxies, ureas, and urethanes. Paints also contain pigments for color. The organic solvents dilute the coating so that it will flow properly upon application. Solvents can be classified by polymer solubility or evaporation rate. For instance, active solvents, typically oxygenated solvents (such as ketones, alcohols, esters, and glycol ethers), usually are miscible with the polymers in all proportions. Diluent solvents (typically hydrocarbons such as toluene, xylene, naphtha, and mineral spirits) have very low polymer solubility, but are used in combination with active solvents to reduce cost. Alternatively, solvents can be grouped by rate of evaporation. To obtain proper atomization and film formation in a spray process, the solvent blend must have solvents with a wide range of evaporation rates. For simplicity, the solvent blend of the spray can be considered to consist of fast- and slow-evaporating solvents. The fast-evaporating solvents give the low viscosity needed for spraying, typically 30 to 100 centipoise (cp), and are about two-thirds of the solvent blend. They must evaporate rapidly from the spray droplets before the droplets impact the surface to be coated. The coating must be deposited with high viscosity, typically 800 to 3000 cp, to keep it from running or sagging. The slow evaporating solvents do not evaporate in the spray; hence, they regulate the coating viscosity during droplet coalescence and film formation. They evaporate (either by air drying or by baking in an oven) as the coating solidifies and forms a smooth and coherent film. This traditional spray method produces high quality coatings that have good appearance and physical properties that are nearly ideal.

Environmental Hazards of VOCs. VOCs are detrimental to the environment in many ways. For example, when released to the atmosphere, they facilitate production of ozone (O_3), a compound which is desirable in the stratosphere to block ultraviolet rays, but which is toxic at ground level. Short-term exposure to ozone contributes to eye, nose, and throat irritation. It also can cause acute lung inflammation and pulmonary edema. Ozone can reduce visual acuity and prolonged ozone exposure can induce emphysema, fibrosis, and bronchopneumonia. It also can lead to congenital birth defects. (1)

VOCs also can migrate from the atmosphere to water systems via precipitation, fallout, or absorption into surface waters. Aquatic VOCs taint seafood, contaminate drinking water, and can be toxic to aquatic life. Moreover, VOCs can penetrate into the soil, where they can injure plant life and inhibit microorganisms responsible for soil fertility (2).

Since the passage of the Clean Air Act (CAA), federal, state, and local regulations have become increasingly stringent concerning VOC emissions. Recently, the CAA regulations have been expanded to include the emission of individual VOCs that are classified as hazardous air pollutants (HAPs). This includes most common coating solvents (particularly those of high volatility) such as methyl ethyl ketone, methanol, hexane, toluene, methyl isobutyl ketone, and xylene. The new regulations impact nearly every type of industrial coating application, including automobiles, trucks and buses, aircraft and marine vessels, wood and metal furniture, appliances and bakeware, machinery and equipment, cans and drums, plastics, consumer goods, building materials, adhesives, and release coatings. Because of the great need to reduce VOC emissions, considerable effort is being put into developing new ways to spray coatings with much less solvent.

Other Coatings Systems Developed to Reduce VOC Emissions

Water-borne coatings use water to replace much of the solvent, but the slow rate of water evaporation and material incompatibility within the coating frequently cause problems in both application and quality. "High-solids" coatings reduce the amount of solvent used by substituting polymers with lower molecular weight for those used in conventional coatings. These low molecular weight polymers then are cured or cross-linked by chemical reaction after being applied. However, high-solids coatings frequently suffer from running, sagging, and cratering; and usually they must be baked at high temperatures. A final example of a new coatings system developed to reduce VOC emissions are powder coatings which are applied as a powder with no solvent and then baked at a very high temperature to melt and fuse the particles into a coherent coating film. Unfortunately, powder coatings generally do not provide the high quality appearance needed for many applications and they usually can not be retrofitted into existing spray lines.

Because of the disadvantages associated with these new coatings systems, low-pollution coating technologies are used only in industries where their characteristics match application requirements, and none of these technologies have generated the general utility or acceptance of conventional solvent-borne coatings. Only recently has a versatile spray system been developed that is able to significantly reduce the emission of VOC solvents and also produce a wide variety of high-quality coatings. This new process uses supercritical carbon dioxide to replace most of the VOC solvents in the spray formulation.

Development of a New Spray System.

A more environmentally attractive spray system has been developed which not only maintains coating quality, but also actually improves coating performance. In commercial applications, this new spray system reduces harmful VOC emissions by up to 80% and can totally eliminate hazardous air pollutants. The new system, called the *supercritical fluid spray process*, uses environmentally-benign supercritical carbon dioxide to replace the fast evaporating solvent used in conventional solvent-borne coatings (3-5). Slow evaporating solvents, which comprise about one third to one fifth of the original solvent blend, are retained to give proper film formation. However, often some of the slow solvent can be eliminated as well because of improved atomization. Furthermore, new reactive liquid polymer systems have been developed that have high carbon dioxide solubility and can be applied as a zero-VOC coatings system.

Supercritical Fluid Spray Process. The supercritical fluid spray process is shown schematically in Figure 1. The reformulated coating, which is called a coating concentrate (removing the solvent gives it high viscosity), is combined with carbon dioxide just prior to spraying. The mixture typically is sprayed at a temperature of about 323 K and pressure of about 100 bar (about 1500 psia). At these conditions, carbon dioxide is a supercritical fluid; that is, the carbon dioxide is above its critical temperature and pressure. However, it should be noted that the polymer / carbon dioxide mixture is below the mixture critical point and is in a single phase region.

Supercritical carbon dioxide is an extremely "fast" evaporating solvent. It dilutes the coating concentrate to a low viscosity for spraying, then evaporates almost instantly after atomization. After the carbon dioxide evaporates, the spray droplets revert to the high viscosity of the coating concentrate, typically 800 to 3000 centipoise (cp). The coating concentrate is formulated (using slow evaporating, active solvents) so that its viscosity is low enough for the coating to coalesce properly on the surface, but high enough to prevent the coating from running or sagging.

Environmental Benefits. The environmental benefits of this supercritical fluid spray technology begin in the workplace. Because fast-evaporating solvents are eliminated and replaced with carbon dioxide, spray booth emissions are reduced to very low levels of VOCs. In fact, the reduction of VOCs in the spray booth can be even greater than the solvent reduction in the overall coating formulation since it is mainly the fast-evaporating solvents which evaporate in the spray booth. Only a small amount of the slow-evaporating solvents evaporate in the spray booth; slow-evaporating solvents evaporate later, during air drying or while baking in an oven. Hence, with the new spray system, workers are exposed to significantly lower concentrations of harmful VOC vapors.

Additional beneficial characteristics of carbon dioxide include its low toxicity and low flammability. While very high concentrations of carbon dioxide can lead to suffocation, the risk is small compared to the toxic effects of organic solvents. This is reflected in its very high threashold-limit value (TLV) of 5000 ppm. This is 10 to 100 times higher than nearly all coating solvents. For example, the TLVs of some common paint solvents are: ethyl acetate (400 ppm), methyl ethyl ketone (200 ppm), methyl isobutyl ketone (100 ppm), toluene (100 ppm), butanol (50 ppm), and hexane (50 ppm). Carbon dioxide also is nonflammable while fast evaporating solvents typically are highly flammable with lower explosive limits of only 1 to 2%. Finally, while fast-evaporating solvents generate most of the odor to which workers are exposed, carbon dioxide is odorless.

Carbon dioxide is the supercritical fluid of choice for several reasons. First, in addition to being nonflammable and odorless, it is noncorrosive and inert. Second, the supercritical state is easily reached because carbon dioxide has a low critical temperature (31 °C; 88 °F), and its critical pressure (7.376 MPa; 74 bar; 1070 psia) is within the operating range of conventional spray equipment. Third, carbon dioxide is inexpensive and readily available in bulk as a byproduct from chemical plants or natural-gas production. Fourth, it has appreciable solubility in coatings even at low solvent levels.

Because the new spray technology directly releases carbon dioxide into the atmosphere, the impact of the technology on the greenhouse effect has come into

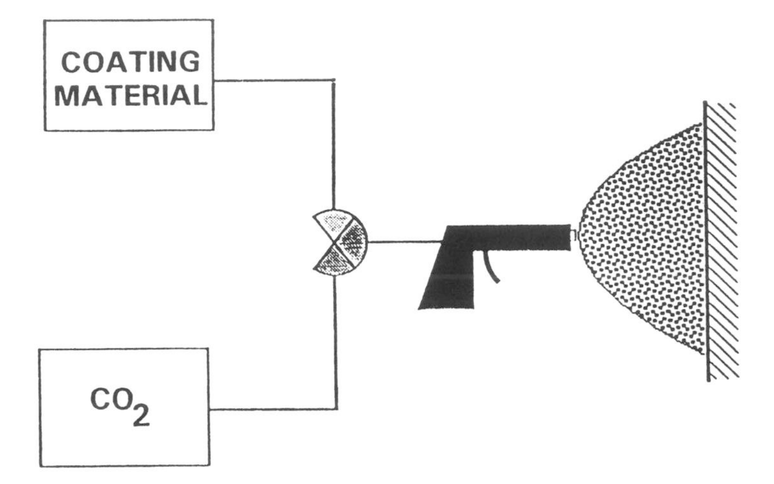

Figure 1: Schematic diagram of the supercritical fluid spray process.

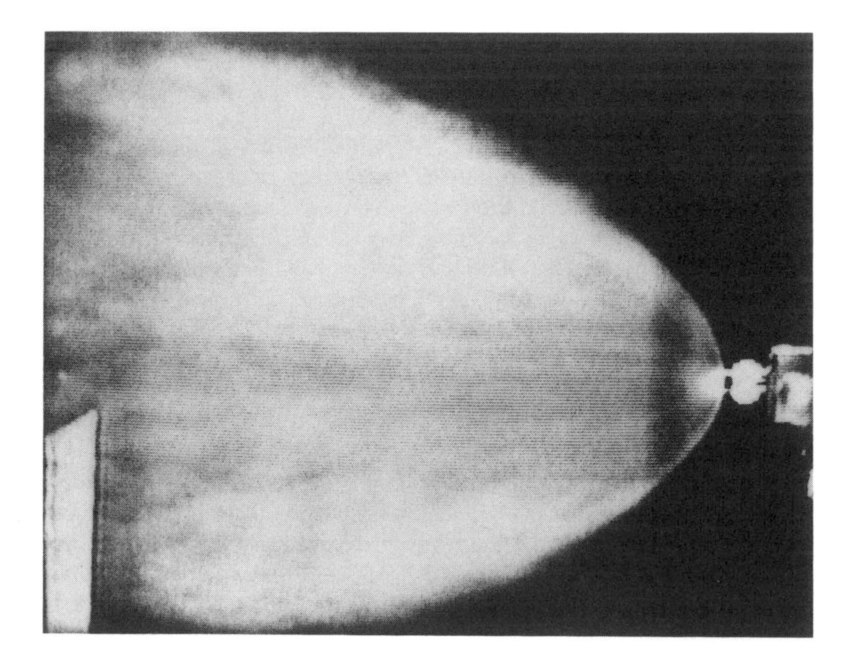

Figure 2: Decompressive spray produced using supercritical fluid spray process.

question. However, the new spray system actually leads to less atmospheric carbon dioxide when compared to conventional spray systems. Organic solvents discharged into the environment from conventional spray systems eventually oxidize in the air to produce carbon dioxide. Each kilogram of solvent emissions from a conventional spray system produces from 2.3 to 3.0 kg of new carbon dioxide. Moreover, if the VOC emissions from a spray booth that uses conventional coatings are incinerated to meet VOC regulations, the natural gas used as fuel for the incinerator can generate up to 18 kg of new carbon dioxide per kilogram of solvent emissions (4). In the new spray system, one kilogram of solvent is replaced by only 0.7 to 1.0 kg of carbon dioxide. Therefore, this new technology actually reduces significantly the total carbon dioxide burden from coating operations.

Atomization. The spray produced by this new technology is very different from conventional pressurized sprays because a different atomization mechanism is involved. In conventional spray processes, atomization occurs by the breakup of a laminar jet due to the development of turbulent instabilities. Solvents are necessary to lower the viscosity of the solid coating before high pressure propels the coating through an orifice at high velocity as a liquid jet. Atomization occurs because the jet becomes unstable from shear with surrounding air; waves grow and break the jet into filaments, which break into droplets. The competing effects of viscosity and surface tension producc a coarse atomization and a non-uniform spray pattern, which limits the quality of the resultant coating.

Use of supercritical carbon dioxide produces a new type of spray, called a decompressive spray, that has superior atomization and spray characteristics compared to conventional spray methods (5,6). As the coating mixture passes through the orifice in the spray nozzle, it depressurizes very rapidly. This depressurization causes the dissolved carbon dioxide to form gas bubbles which reduce the speed of sound in the flow, and hence retard the rate that the pressure drops, until a condition of "choked flow" is reached (7). Choked flow occurs when the flow rate through the orifice reaches the speed of sound. Because the pressure drop can not propagate faster than the speed of sound, this choked flow causes a high pressure zone to extend beyond the orifice. The very rapid decompression (and evolution of the carbon dioxide vapor that follows) creates intense shear and a powerful expansive force which overcome the liquid forces of viscosity, cohesion, and surface tension. The expanding carbon dioxide from the high pressure zone outside of the orifice ejects droplets outward in all directions, so the spray becomes parabolic with an angle at the nozzle that is nearly 180°. The parabolic shape is due to the interaction between the outward decompression and the high forward momentum of the spray. This is shown in Figure 2.

The spray has a "feathered" pattern with a uniform interior and tapered edges which is desirable for overlapping the spray to form coatings with uniform thickness (8). Moreover, the parabolic spray can be constrained into forming a two-dimensional planar spray pattern if a v-shaped groove is cut at the orifice outlet. The droplet size is also very uniform everywhere in the spray pattern. Therefore, decompressive atomization substantially reduces the number of large droplets (which are detrimental to coating appearance) and the number of very small droplets (which waste coating, because they are blown past the article being sprayed).

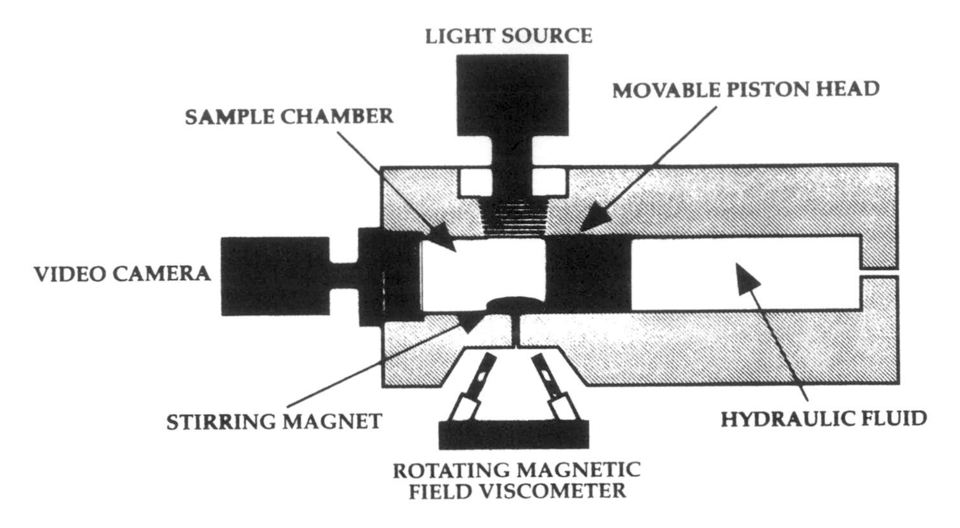

Figure 3: Phaser apparatus for measuring phase boundries and viscosity.

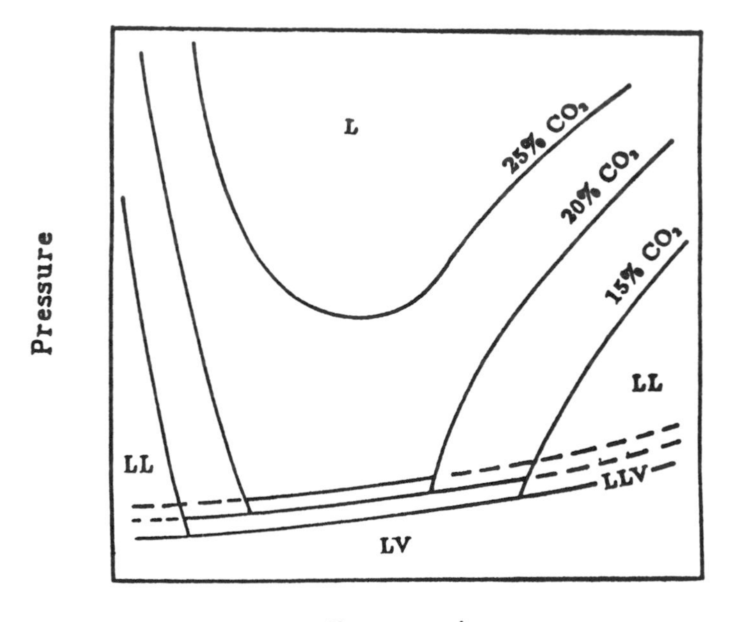

Figure 4: Schematic diagram showing the pressure-temperature behavior for a carbon dioxide / solvent / polymer mixture as a function of carbon dioxide concentration. The diagram shows both the low-temperature (UCST) and the high-temperature (LCST) regions of liquid-liquid immiscibility.

Thermodynamics. To better understand carbon dioxide solubility in coatings, and therefore the spray conditions used in the supercritical fluid spray process, it is helpful to understand the thermodynamics of polymer-solvent-carbon dioxide mixtures in general. The phase relationships and physical properties have been measured using a variable-volume, high-pressure view cell, which is called a Phaser (see Figure 3). The Phaser is a modification of an experimental apparatus designed by McHugh and co-workers (9-11) and was made by Union Carbide Corporation. It is capable of measuring density, viscosity, solubility, and pressure-temperature phase boundaries. The Phaser consists of a cylinder with a movable piston that partitions the central bore into a sample chamber and a hydraulic fluid chamber. A light source and video camera are attached to the sample chamber to view the phase condition, and the contents of the sample chamber are mixed by a magnetic stirrer. To obtain data, a measured sample of polymer and solvent, which may contain 0 to 100% polymer, is loaded into the sample chamber, and then a measured amount of carbon dioxide is added. As the mixture is stirred, the temperature is adjusted by electrical resistance heating, and the pressure is adjusted by a pressure generator that pressurizes the hydraulic fluid. When the mixture reaches equilibrium, its viscosity is measured by a rotating magnetic field viscometer and the phase condition is observed. Solubility limits and phase boundaries for the ternary system are identified by changing the pressure and temperature and observing the phase changes at different carbon dioxide levels.

The Phaser experiments reveal a surprising thermodynamic principle of polymer-solvent-carbon dioxide mixtures. Whereas most polymers have very low solubility (often below 1%) in supercritical carbon dioxide, supercritical carbon dioxide can have appreciable solubility in many types of polymers. Carbon dioxide solubilities ranging from 20 to 60% have been measured for several pure liquid polymer systems having no organic solvent. In these systems, the thermodynamic concepts of solvent and solute are reversed; the carbon dioxide is the solute and the polymer is the solvent.

This principle allows the supercritical fluid spray process to be used with coatings having very high polymer levels and only enough slow solvent to give the proper viscosity for coalescence and film formation. The only diluent is carbon dioxide, which provides low viscosity for spraying and forms the decompressive spray. The process also can be used with low molecular weight reactive (two-component) systems that have no solvent. For these systems, the viscosity increase that normally results from solvent evaporation is accomplished by using catalysts, heat, or ultraviolet light to initiate chemical reactions which cross-link and solidify the coating.

Phase Chemistry. The miscibility of supercritical carbon dioxide with solutions of polymer and organic solvent, in general, depends upon the carbon dioxide level, temperature, pressure, polymer level, and compatibility of the polymer with the solvent. As the carbon dioxide level in the ternary mixture is increased (for a fixed ratio of polymer to solvent), higher pressure is required to obtain complete miscibility, and it is found that there is a narrower temperature range for the region of miscibility (12,13), i.e., a single-phase liquid solution (L). This is illustrated schematically in Figure 4. The

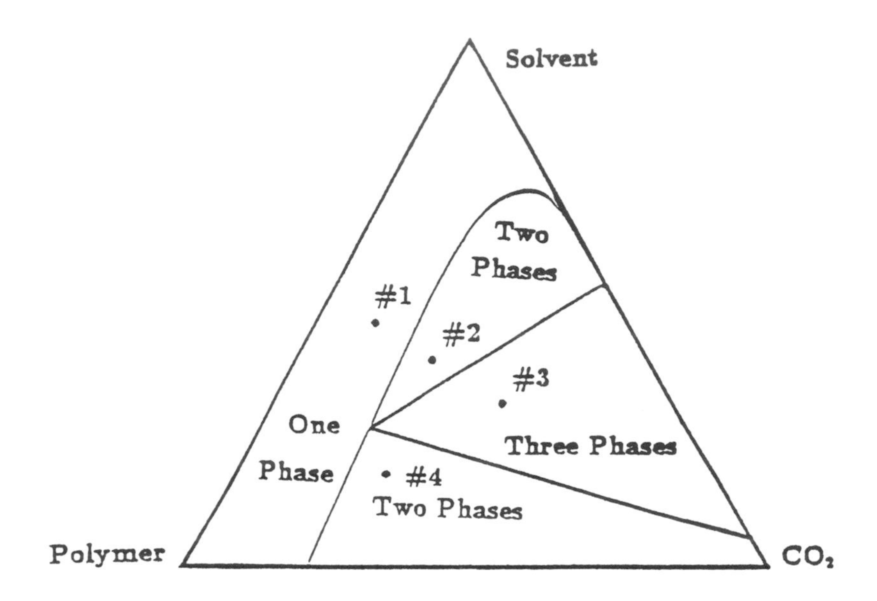

Figure 5: A schematic ternary diagram for a carbon dioxide / solvent / polymer mixture.

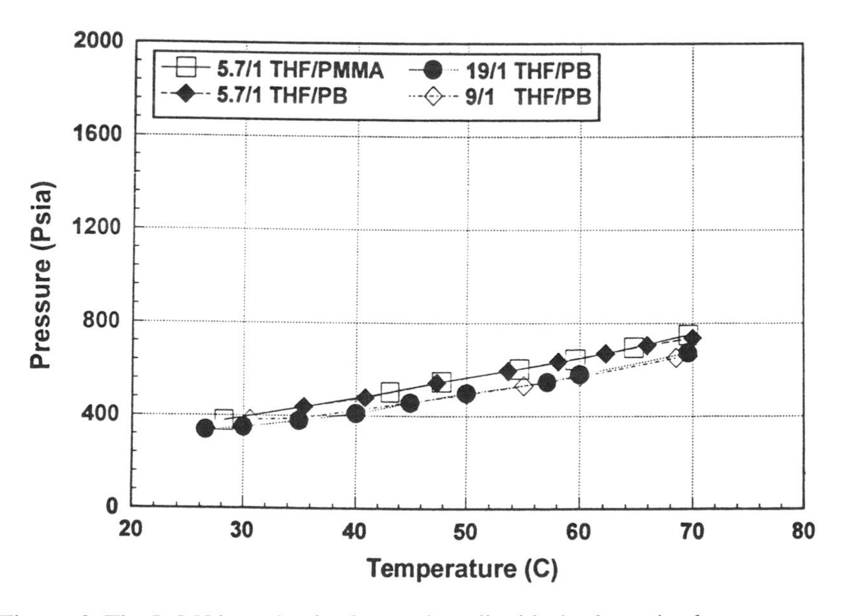

Figure 6: The L-LV boundry for four carbon dioxide / solvent / polymer systems. Systems with different types of polymer and varying solvent to polymer ratios are shown. All systems contain 30% (by weight) carbon dioxide.

phase diagram shows the boundaries between regions having different number and types of phases. The boundary between the single liquid (L) and liquid-vapor (LV) regions is called the bubble-point line, because it is the point where vapor bubbles first form during depressurization. The boundaries between the liquid (L) and liquid-liquid (LL) regions are called cloud-point lines, because they are the points at which liquid droplets of a second liquid phase first appear during depressurization. This usually is seen as a sudden transition from a clear solution to a white cloudy mixture. The cloudy appearance is caused by separation of the single liquid into two immiscible liquid phases.

The L-LL boundary has two sections: the low-temperature boundary (left side) is called the Upper Critical Solution Temperature (UCST) line and the high-temperature boundary (right side) is called the Lower Critical Solution Temperature (LCST) line. (It should be noted that the Lower Critical Solution Temperature usually is above the Upper Critical Solution Temperature.) As the carbon dioxide level increases, the UCST line moves to higher temperatures and pressures, and the LCST line moves to lower temperatures and higher pressures. The bubble-point line moves to higher pressures. Consequently, the single-liquid region of complete miscibility shrinks. Ultimately, the LCST and UCST lines merge into a single line at a high carbon dioxide level. Then the liquid-liquid region separates the liquid and liquid-vapor regions (11-13). Generally, coatings are sprayed within the liquid region close to the LCST boundary.

The phase boundaries for different compositions of polymer, solvent, and carbon dioxide by weight, at constant temperature and pressure, are shown by the triangular diagram illustrated in Figure 5. The corners of the triangle are the pure components and each side gives compositions with just two components. The perpendicular distances from any point to the sides give the weight fractions of each component in the overall composition. At low polymer levels, adding carbon dioxide moves the mixture from the single-phase region (point #1), into the liquid-liquid region (point #2), and then into the liquid-liquid-vapor region (point #3), where the vapor phase typically contains more than 90% carbon dioxide and very little polymer. For medium to high polymer levels, adding carbon dioxide moves the mixture into a liquid-vapor area (point #4). At higher pressure, the two-phase and three-phase regions can merge into a single liquid-liquid region.

The effect of polymer type and polymer level on the phase boundaries can be shown on a pressure-temperature diagram by keeping the carbon dioxide level and solvent type constant. For low to medium polymer levels, experiments show that the polymer type and polymer level have little effect on the L-LV boundary (bubble-point line) (16). This is illustrated in Figure 6 for four systems having 30% carbon dioxide and tetrahydrofuran (THF) as the solvent. The following polymers and solvent/polymer ratios were used: polybutadiene (PB) at 19/1, 9/1, 5.7/1 as well as with polymethyl methacrylate (PMMA) at 5.7/1. The L-LV boundaries for each system have the same slope and nearly overlap.

In contrast, measurements show that changing the polymer level has a large effect on the L-LL boundary (cloud-point line). This is illustrated in Figure 7 for three systems having 23% carbon dioxide and tetrahydrofuran/polystyrene ratios of 5.7/1, 4/1, and 3/1. The curves for 5.7/1 and 4/1 ratios exhibit the generally observed trend that as the

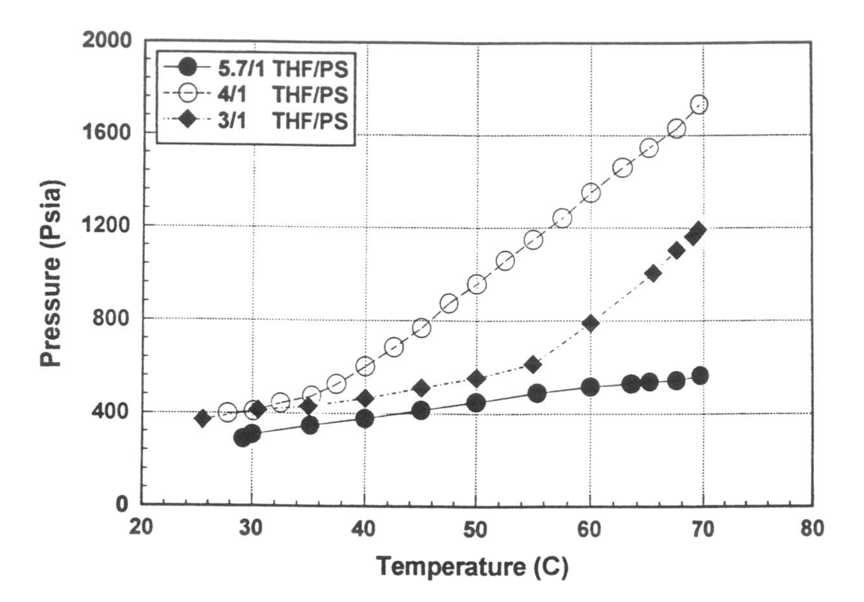

Figure 7: The L-LL boundry for three systems with varying solvent to polymer ratios. Each system contains 23.1% (by weight) carbon dioxide.

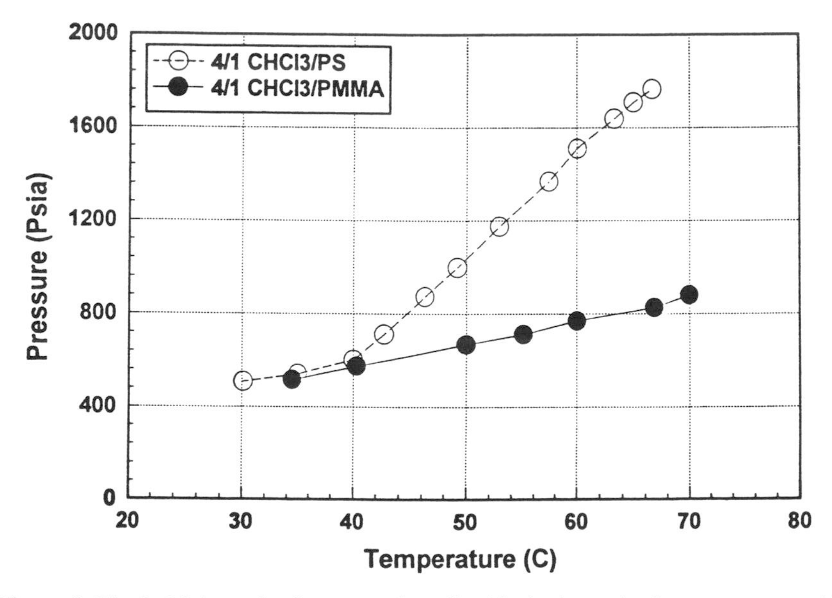

Figure 8: The L-LL boundry for two carbon dioxide / solvent / polymer systems with different types of polymers.

polymer level increases, the L-LV boundary shifts to higher pressures and lower temperatures. However, the curve for the higher polymer level with a 3/1 ratio does not follow this; it reverses the trend.

Changing the polymer type also can have a large effect on the L-LL boundary. This is illustrated in Figure 8 for two systems having the same carbon dioxide levels (20%), the same solvent (chloroform), and a 4/1 ratio of solvent to polymer, but with either polymethyl methacrylate or polystyrene as the polymer. The L-LL boundary for PS begins at about 40 C, but the L-LL boundry for PMMA is above 70 C and could not be measured. It should be noted that the opposite trend is seen if the solvent is tetrahydrofuran (i.e. the L-LL boundry for PMMA is at a lower temperature). In general, changing the solvent changes the phase boundaries and the L-LV boundary (bubble-point line) usually is affected more by changing the solvent than by changing the polymer type or polymer level.

The L-LL boundary (cloud-point line) also varies considerably with solvent. This is illustrated in Figure 9 for three systems having 25% carbon dioxide, polymethyl methacrylate polymer, and a solvent/polymer ratio of 3/1, but having methyl ethyl ketone, acetone, or tetrahydrofuran as the solvent. The L-LL boundaries are very different; the liquid-liquid region extends to the lowest temperatures for methyl ethyl ketone and is at the highest temperatures for tetrahydrofuran.

These thermodynamic data illustrate general solubility trends that need to be considered when using supercritical carbon dioxide to spray coatings. Commercial coatings usually have even higher polymer levels and may use different polymers and solvents than those used in these studies. Measurements of the bubble-point and cloud-point curves for a coating help to establish the temperature, pressure, and carbon dioxide level to use for spraying. Different coatings are generally sprayed at different conditions. In general, spray temperatures range from 35 to 65 °C and spray pressures range from 80 to 135 bar. Further thermodynamic research is needed to obtain a complete understanding of the phase behavior of polymer-solvent-carbon dioxide systems, especially at higher polymer levels and for commercially important coating systems.

Viscosity. Supercritical carbon dioxide is more effective than most liquid solvents in decreasing the viscosity of a polymer. A comparison of the viscosity-reducing potential of carbon dioxide versus tetrahydrofuran is shown in Figure 10 for the polymer, polyisobutylene. As indicated by the figure, when the weight percentage of polymer is fixed, carbon dioxide will create a mixture viscosity much lower than that created by tetrahydrofuran.

Unlike liquid solvents, supercritical carbon dioxide does much more than dilute the coating to a low spray viscosity -- it changes the nature of the atomization mechanism. Thus, low spray viscosity is less important than with conventional spray methods, and many new factors contribute to the spray-ability of the coating.

Application of Phase Chemistry to the New Spray Process. In order to form a decompressive spray and to use solvent efficiently to form high quality coatings, the

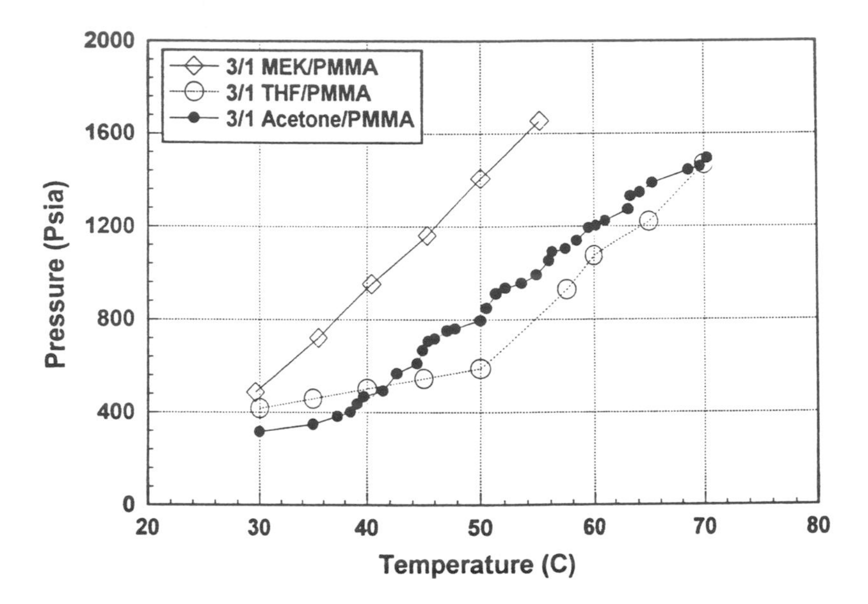

Figure 9: The L-LL boundry for three carbon dioxide / solvent / polymer systems with different solvents. Each system has a 3/1 solvent to polymer ratio and a carbon dioxide concentration of 25 %.

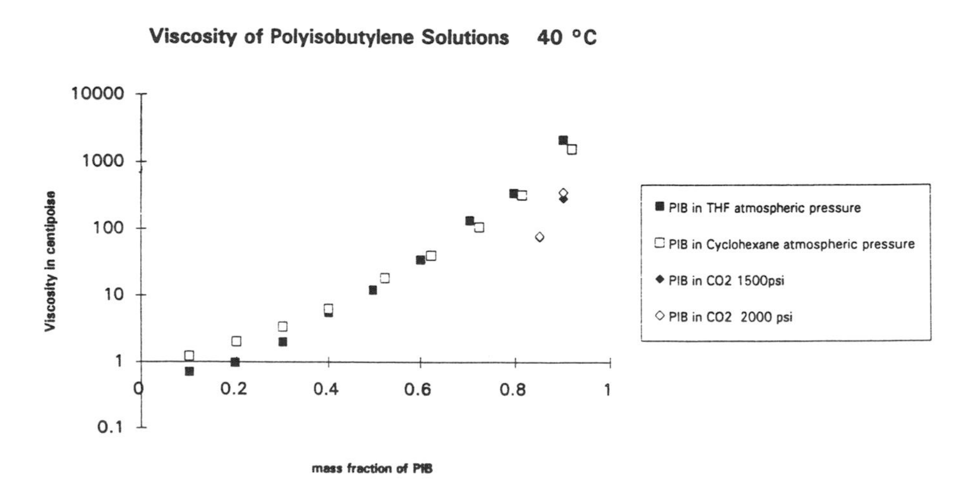

Figure 10: Viscosity of polyisobutylene solutions at 40°C. Data are presented for PIB in tetrahydrofuran, cyclohexane, and supercritical carbon dioxide.

mixture of coating and carbon dioxide must not only have low viscosity, it must be sprayed at proper combinations of temperature, pressure, and concentration of carbon dioxide. The most desirable spray conditions occur when the carbon dioxide is completely dissolved in the coating as a single-phase solution that is close to the solubility limit or phase boundary for the system. This condition promotes formation of carbon dioxide gas during spraying and produces the greatest expansive force for atomization.

During depressurization, the carbon dioxide gas phase can be formed from the dissolved carbon dioxide either directly or indirectly, depending upon the relationship between the spray conditions and the phase boundaries for the system. In the first case, the gas phase is formed directly as the liquid spray solution drops below the bubble-point pressure. This creates a liquid-vapor spray mixture that expands in gas volume as the mixture decompresses (depressurizes). In the second case, the dissolved carbon dioxide first forms a liquid phase, which then forms the gas phase as the spray mixture drops below the bubble-point pressure. The mixture begins in the liquid phase, transforms into a liquid-liquid mixture, then becomes a liquid-liquid-vapor mixture, and finally exists as a liquid-vapor mixture. In this process, the mixture crosses three phase boundaries, not just one. Although both processes are effective in producing a decompressive spray, the second case typically gives finer atomization. This is believed to occur because first forming the liquid carbon dioxide phase and then the gas phase produces more, but smaller, bubbles than when the gas phase is formed directly. The higher concentration of gas bubbles generates a more uniform atomization.

Conclusions

VOC emissions can be reduced during the spraying of coatings by replacing liquid solvents with supercritical carbon dioxide. There are both health and environmental benefits of this new process. Worker exposure to VOCs in the spray booth can be reduced significantly, and the level of harmful ozone generated in ground level air is reduced. The new system actually reduces the greenhouse effect, by directly releasing carbon dioxide into the atmosphere at a rate much lower than the rate that the VOC emissions of a conventional system produce carbon dioxide in the atmosphere.

Unlike previously developed, low-pollution spray systems, the coatings produced by the supercritical carbon dioxide spray are of high quality and thus are highly marketable. Due to the unique decompressive atomization mechanism caused by choked flow, the new spray has a feathered pattern with a narrow droplet-size distribution which produces coatings of more uniform thickness than those of conventional sprays.

Spray and phase behavior studies have shown that optimal atomization occurs when the miscible liquid spray passes through a liquid-liquid region and then a liquid-liquid-vapor region before becoming a liquid-vapor mixture as it decompresses. Thermodynamic studies have focused on the miscibility of the polymer-solvent-carbon dioxide mixtures and revealed that carbon dioxide actually functions as the solute and the polymer as the solvent in these systems.

Acknowledgments

We gratefully acknowledge support of this research by the National Science Foundation and Union Carbide Corporation. Financial support was provided by the Chemical and Thermal Systems Division of NSF (contract number CTS-9216923) under their program for Environmentally Benign Chemical Processing. The equipment for the experiments was loaned to The Johns Hopkins University by Union Carbide Corporation.

We thank Lisa Smith and Rebecca LePosa for collecting viscosity data. Funding of Lisa Smith, Rebecca LePosa, and Jennifer Geiger was provided by an NSF supplement for Research Experiences for Undergraduates.

Literature Cited

(1) Suess, M. J.; Grefen, K.; Reinisch, D. W. *Ambient Air Pollutants From Industrial Sources*; Elsevier: New York, 1985; pp 65-73.

(2) National Research Council, Division of Medical Sciences Assembly of Life Sciences. In *Vapor-Phase Organic Pollutants*; National Academy of Sciences: Washington, D.C., 1976; pp 236-270.

(3) Lee, C.; Hoy, K. L.; Donohue, M. D. Supercritical Fluids as Diluents in Liquid Spray Application of Coatings. U.S. Patent 4 923 720, 1990; U.S. Patent 5 027 742, 1991.

(4) Nielsen, K. A.; Busby, D. C.; Glancy, C. W.; Hoy, K. L.; Kuo, A. C.; Lee, C. Supercritical Fluid Spray Application Technology: A Pollution Prevention Technology for the Future. In *Proceedings of the Seventeenth Water-Borne and Higher-Solids Coatings Symposium*; Storey, R. F.; Thames, S. F., Eds.; University of Southern Mississippi: Hattiesburg, MS, 1990; pp 218-239.

(5) Hoy, K. L.; Nielsen, K. A.; Lee, C. Liquid Spray Application of Coatings with Supercritical Fluids as Diluents and Spraying from an Orifice. U.S. Patent 5 108 799, 1992; U.S. Patent 5 203 843, 1993.

(6) Nielsen, K. A.; Glancy, C. W.; Hoy, K. L.; Perry, K. M. A New Atomization Mechanism for Airless Spraying: The Supercritical Fluid Spray Process. In *Proceedings of the Fifth International Conference on Liquid Atomization and Spray Systems*; Semerjian, H. G., Ed.; NIST Pub. 813: Gaithersburg, MD, 1991; pp 367-374.

(7) Colwell, J. D.; Senser, D. W.; Nielsen, K. A. Influence of Temperature on the Structure of Supercritical Fluid Coating Sprays. In *Proceedings of the Sixth Annual Conference on Liquid Atomization and Spray Systems*; Peters, J., Ed.; University of Illinois: Urbana, IL, 1993; pp 39-43.

(8) Nielsen, K. A.; Hoy, K. L.; Bok, H. F. Methods and Apparatus for Obtaining a Feathered Spray When Spraying Liquids by Airless Techniques. U.S. Patent 5 057 342, 1991; U.S. Patent 5 141 156, 1992.

(9) Seckner, A. J.; McClellan, A. K.; McHugh, M. A. *AIChE J.* 1988, *34*, 9.

(10) McClellan, A. K.; Bauman, E. G.; McHugh, M. A. Polymer Solution Supercritical Fluid Phase Behavior. In *Symposium Proceedings on Supercritical Fluid Technology*; Penninger, J. M. L.; Rodosz, M.; McHugh, M. A.; Krukonis, V. J., Eds.; Elsevier Science: New York, 1985; pp 161-178.
(11) McHugh, M. A.; Guckes, T. L. *Macromolecules* 1985, *18*, 674.
(12) Kiamos, A.A., High-Pressure Phase-Equilibrium Studies of Polymer-Solvent-Supercritical Fluid Mixtures, Masters Essay, The Johns Hopkins University, 1992.
(13) Kiamos, A.; Donohue, M. D. The Effect of Supercritical Carbon Dioxide on Polymer-Solvent Mixtures, *Macromolecules* 1994, 27, 357.

RECEIVED December 29, 1995

Chapter 13

Chemically Benign Synthesis at Organic–Water Interface

Phooi K. Lim and Yaping Zhong

Department of Chemical Engineering, North Carolina State University, Raleigh, NC 27695–7905

An organic-water interfacial synthesis technique has been developed which is based on the use of a surface-active catalyst complex in conjunction with an aqueous organic solvent mixture and an emulsifier to effect the desired reaction at the organic-water interface. By decoupling the various functions of the reaction medium and meeting them separately with a combination of aqueous and organic phases, instead of a single organic solvent, chances are improved for an environmentally benign synthesis. The technique also offers other significant advantages, including the ease of catalyst recovery and processing, a high reactivity, selectivity and reproducibility under mild reaction conditions, and a greater versatility compared to phase-transfer, micellar and other biphasic techniques. Results of the application of the technique to several reactions of commercial and scientific interests are presented.

Environmental Rationale for an Organic-Water Interfacial Synthesis

In a conventional liquid-phase synthesis, the solvent--typically an organic liquid--can take on multiple functions that may include (1) solubilizing not only the reactant(s), but also any catalyst or intermediate product(s) of the reaction, (2) serving as a catalyst ligand and promoting the desired reaction through an electronic or steric effect at the catalyst, and (3) acting as an acid or base. The first function is by itself restrictive on the choice of the solvent because an organic substrate tends to be oleophilic whereas a metal catalyst tends to be hydrophilic. If two or more reactants are involved, a similar solubility incompatibility problem may also exist, with one reactant being hydrophilic and the other(s), oleophilic. If additional functional requirements are added to the first, the choice of the solvent will quickly become

0097–6156/96/0626–0168$12.00/0

severely limited. The solvent candidates left in the pool will likely be toxic, hazardous, or environmentally troublesome. For example, pyridine has been found to be a very good solvent for the coupling polymerization of 2,6-dimethylphenol (*1-3*) because it fulfills all the three functions mentioned above. Unfortunately, it is also highly toxic and hazardous.

By removing the somewhat arbitrary constraint of a single phase, one can decouple the various functions that limit the choice of a solvent and fulfill them separately by a combination of aqueous and organic phases. With fewer functional requirements to satisfy, the organic component in the biphasic mixture can be chosen from a much wider field, and chances are greater that a more benign organic liquid can be found.

The premise that a biphasic mixture can fulfill the requirements of a liquid-phase reaction system while at same time be environmentally benign is the basis of an organic-water interfacial reaction technique which has been developed at North Carolina State University (*4-6*). An aqueous solution and a relatively innocuous organic liquid are used to dissolve the catalyst and reactant in their respective phases. A surface-active complexing agent or ligand is used to draw the catalyst to the organic-water interface where it effects the desired reaction. An emulsifier is used to promote phase dispersion and increase the interfacial area for the reaction. The interfacial technique is different from, though complementary to, the phase-transfer and micellar techniques and the other biphasic technique which others have developed based on the use of water-soluble ligands (see later).

Toluene has been found to serve well as the organic phase in many biphasic reactions. Although it is not completely innocuous, it is much safer and easier to handle than many of the organic solvents which have been used in conventional syntheses, including pyridine, nitrobenzene, dimethylformamide, dimethylsulfoxide, acetonitrile, and acetic acid. Moreover, toluene can be easily recovered and recycled. Thus, any risk presented by the use of toluene can be reduced by recycling and by the knowledge and experience which have been acquired on its safe handling.

The amount of surface-active ligand needed is typically one to three times the amount of the catalyst, and, as will be shown later, the complexing agent and the catalyst, as well as the organic solvent, can often be recovered easily and recycled. The amount of emulsifier needed is typically less than 0.010 M, and it, too, can be recycled. Any safety hazards posed by the complexing agent and the emulsifier are minimized by their low concentrations and can be further reduced by recycling.

Other Possible Advantages of Organic-Water Interfacial Synthesis

Aside from the environmental benefits, the organic-water interfacial technique also offers some other significant advantages, including (1) ease of catalyst recovery and reuse, and ease of operation and control; (2) high reactivity, selectivity, and reproducibility under mild reaction conditions (features that are normally associated with a homogeneously-catalyzed reaction system); (3) a higher solubility for a gaseous reactant than in an aqueous reaction system; (4) a higher reaction rate due to the concentrating and intimate-contacting effects of the interface on the catalyst and reactant(s); and (5) possibility of some stereoselectivity control due to the directional influence of the interface on molecular orientations.

The ease of catalyst recovery arises from the surface activity of the catalyst complex and the ability of the biphasic reaction mixture to phase-separate upon standing. The catalyst complex can be found mostly in a compact emulsion layer between the organic and aqueous phases; it can easily be recovered and reused.

While drawn by surface forces to the interface, the catalyst complex is still free to take up a configuration that is most favorable to the reaction. Thus, compared to a solid-supported catalyst, it will function more like a homogeneous catalyst and show more of the latter's characteristic features, namely, high reactivity, selectivity, and reproducibility under mild reaction conditions. Moreover, because the catalyst complex is in a dissolved state, the reaction mixture may also be easier to handle than the slurry counterpart that would result from the use of a solid-supported catalyst.

The presence of an organic phase can significantly increase the solubility of a gaseous reactant because organic liquids generally have higher solubilities for gases than water. For example, on the same volume basis, toluene can dissolve 7.2, 6.7, and 48.2 times as much of carbon monoxide, oxygen, and hydrogen, respectively, as water (*7,8*). For heptane, the ratios are even higher, being 11.1, 11.5, and 120 (*7,8*), respectively. Thus, the use of a 1:1 aqueous organic mixture can increase the gas solubility by a factor of 3 to 60 over the aqueous system. The higher gas solubility will reduce the chances of the reaction being limited by mass transfer. It can also increase safety margin by reducing the pressure of a gaseous reactant that is either toxic, e.g., carbon monoxide, or hazardous, e.g., hydrogen.

Finally, the organic-water interface has the ability to attract, collect, and concentrate a reactant, if the latter has some amphiphilic character. The concentrating effect and the resulting intimate contact between the reactant and catalyst at the interface are favorable to a fast reaction.

Apparatus and Procedure in the Interfacial Synthesis Development

Apparatus. The interfacial synthesis reactions described below are each carried out batchwise under mild reaction conditions (<70°C and 1 atm) in a three-necked, 585-mL Morton-flask reactor that has the following provisions (*4-6*): (1) a Teflon seal joint that permits the insertion and free rotation of the shaft of a mechanical stirrer through the middle neck of the reactor while keeping the reactor air-tight; (2) introduction of the biphasic reaction mixture and delivery of a gaseous reactant through a stoppered side neck; (3) catalyst addition through the other, septumed side neck; and (4) temperature control by a thermostated water bath.

The Teflon seal joint (Fisher Scientific catalog number 14-513-100) consists a Teflon bearing with an exterior 24/40 taper, an O-ring, and an adapter chuck. The septumed side neck carries a needle-stemmed spoon that can be lowered into the reactor solution to unload the catalyst it carries. The other side neck is connected through a three-way valve to a gas tank and a constant-pressure manometric unit. The gas tank supplies the gaseous reactant while the manometric unit provides a convenient and sensitive method for monitoring the synthesis reaction by means of the gas uptake. The manometric unit (*4-6*) consists of two mobile arms of equal-diameter burets attached to an adjustable pulley system that is actuated by two flywheels. The burets are interconnected by a flexible tubing at their bottoms, and they are confined--by cords winding through the flywheels--to move up and down in countervailing unison so that a constant manometer head is maintained. The burets are partially filled with water, which serves as the manometer fluid. During the reaction, the relative height of the burets is adjusted continuously to give a constant manometer head. The gas uptake reading is registered by the liquid level in the buret directly hooked up to the reactor.

Procedure. A typical procedure is as follows (*4-6*): The reactor is charged with 150 mL of a 1:1 (by volume) aqueous-organic mixture and the desired amounts of substrate, emulsifier, surface-active ligand, and any base (needed for carbonylation and oxidative coupling reactions). The desired amount of catalyst is placed on the needle-stemmed spoon which is raised to a suspended position above the reactor solution. The reactor is placed in a water bath and purged with the gaseous reactant for about 10 minutes and then allowed to equilibrate to the set temperature. The synthesis reaction is initiated by lowering the needle-stemmed spoon into the reactor solution to unload the catalyst on it. The stirrer is turned on, and the gas uptake is followed manometrically.

Upon completion of the reaction, the reaction mixture is allowed to phase separate by standing. The catalyst and the surface-active ligand are recovered mostly from the interface in the form of a compact emulsion and suspension. The main product is recovered from the organic or aqueous phase by acidification and purified by recrystallization; its identity is confirmed by comparison with a true standard using proton-NMR, FTIR, GC-MS and UV-visible spectroscopies. The catalyst distribution in the biphasic mixture is determined by ICP (or inductively-coupled plasma atomic emission) spectroscopy, and the activity of the recovered catalyst complex is confirmed by repeating a run. The organic phase can also be recycled, after a distillation step to remove the residual product and the acid and alcohol components introduced in the acidification step. Molecular weight of a polymer product is determined by intrinsic viscosity measurements and gel permeation chromatography.

Specific Applications of the Organic-Water Interfacial Technique

Coupling Polymerization of 2,6-Dimethylphenol. Reaction (1) is of interest

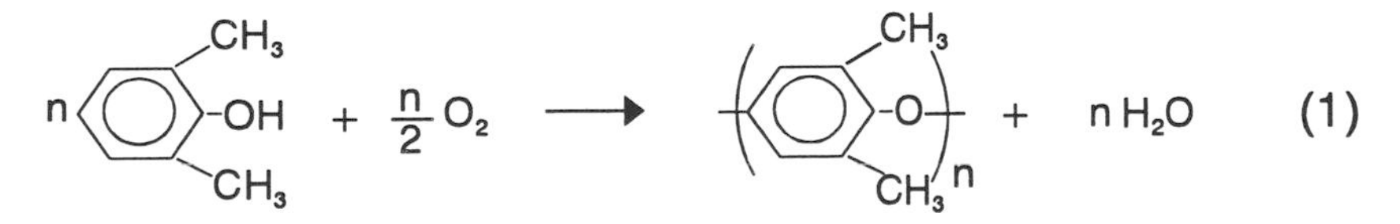

because poly(phenylene oxide), the reaction product, is a commercially-important thermoplastic resin with many applications (*4*). High molecular-weight (>80,000) polymers have been obtained in a high yield (>95%) at 25°C and 1 atm in less than two hours using the interfacial technique (*4,5*).

In the interfacial synthesis, toluene and aqueous ammonia are used as the biphasic reaction medium, cuprous-chloride-triethylphosphite complex as the surface-active catalyst complex, and dodecyl sodium sulfate (DSS) as the emulsifier. Other organic liquids, e.g., chloroform, may be used in place of toluene, but they offer no significant advantage. Other surface-active ligands are less effective than triethylphosphite. In the absence of the latter, little or no polymer product is formed. Cationic and nonionic surfactants are less effective than DSS as emulsifiers.

The reaction shows a strong dependence on stirring. The optimal organic-water phase ratio is around 1:1 (by volume) and the optimal aqueous ammonia concentration is 0.26 M. The reaction orders with respect to catalyst, dimethylphenol, and oxygen are 2.0, 1.0, and 0.15, respectively. The activation energy of the reaction is 14.7 kcal/mol. The reaction mixture undergoes a spontaneous phase separation upon standing, and the catalyst complex can be recovered and reused, with no apparent loss of activity or selectivity.

The interfacial synthesis compares favorably with the conventional homogeneous synthesis that makes use of solvents such as pyridine, nitrobenzene, chlorobenzene, or mixtures of organic amines with toluene (*1-3,9-12*). The formation rate, yield, and molecular weight of the polymer product are all comparable to that of the conventional homogeneous synthesis.

Carbonylation of Benzyl Chloride. Reaction (2) is of interest because phenylacetic

$$C_6H_5-CH_2Cl + CO + H_2O \longrightarrow C_6H_5-CH_2COOH + HCl \qquad (2)$$

acid, the reaction product, is used as a perfume and flavor additive and as a feed component in the enzymatic production of Penicillin G. The interfacial technique has been applied to the reaction, and quantitative yield of phenylacetic acid has been obtained at 60°C and 1 atm in less than three hours (*6*).

In the interfacial synthesis, toluene and aqueous sodium hydroxide are used as the biphasic reaction medium, palladium-(4-dimethylaminophenyl) diphenylphospine complex as the surface-active catalyst complex, and DSS as the emulsifier. The nature of the phosphine ligand is important. Little or no reaction occurs with triethoxylphenylphosphine and triphenylphosphine as the ligand. Presumably, (4-dimethylaminophenyl)diphenylphosphine is effective in promoting the reaction because it is surface-active. A cationic surfactant, e.g., cetyltrimethylammonium bromide, may be used in place of DSS as the emulsifier. In that event, however, the surfactant also functions as a phase-transfer agent, and the overall reaction has a phase-transfer component in addition to the interfacial component.

Palladium can be recovered quantitatively at the organic-water interface after the reaction, and it shows the same catalytic activity as the fresh catalyst under similar conditions. The reaction has an activation energy of 23.9 kcal/mol and a first-order dependence on the catalyst and substrate concentrations each, a zero-order dependence on the carbon monoxide partial pressure, and a variable-order dependence on the aqueous sodium hydroxide concentration.

Carbonylation-Induced Cross Coupling of Diamino- and Dihalo-Aromatics. Reactions (3) and (4) typify a large class of carbonylation-induced cross coupling reactions which can be effected using the interfacial technique. The reactions were originally developed by Perry (*13*) using a homogeneous technique that makes use of dimethylacetamide and 1,8-diazabicyclo[5.4.0]undec-7-ene as the organic solvent

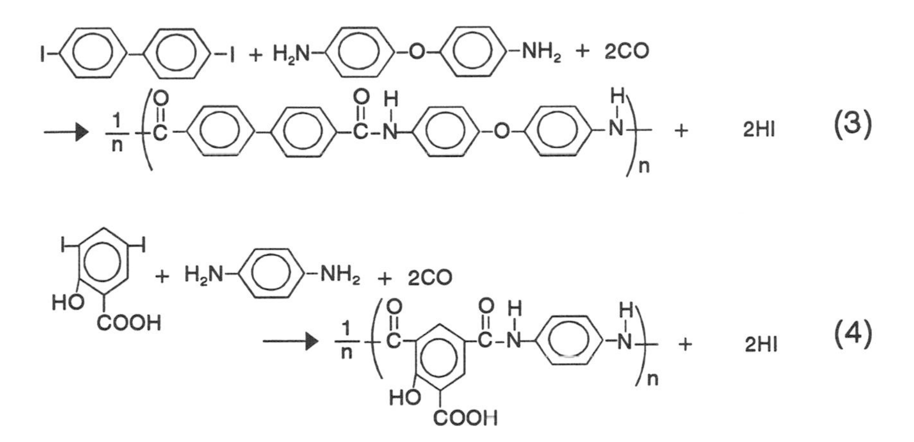

and base, respectively. The reactions are of interest because they provide a general avenue for the synthesis of a broad range of aromatic polyamides or aramid polymers. In principle, different combinations of diamino- and dihalo-aromatics can be used to produce aramid polymers with different properties. Thus, while reaction (3) produces a polymer that is insoluble in most solvents except concentrated sulfuric acid, reaction (4) produces a polymer that is soluble in an aqueous base solution.

The catalyst, reagents, and reaction conditions used to effect the interfacial carbonylation of benzyl chloride can also be used to effect the interfacial carbonylation-cross coupling reactions. Triethoxylphenylphosphine, however, has been found to be a more effective catalyst ligand than (4-dimethylaminophenyl)diphenylphosphine in these carbonylation-cross coupling reactions. The results obtained to date are encouraging. Additional kinetic and characterization studies are being carried out; the results will be reported at a later date.

Autoxidation of Tetralin. The interfacial technique has also been applied to the autoxidation of tetralin (Lim *et al.*, submitted to *J. Phys. Chem.*). The reaction is

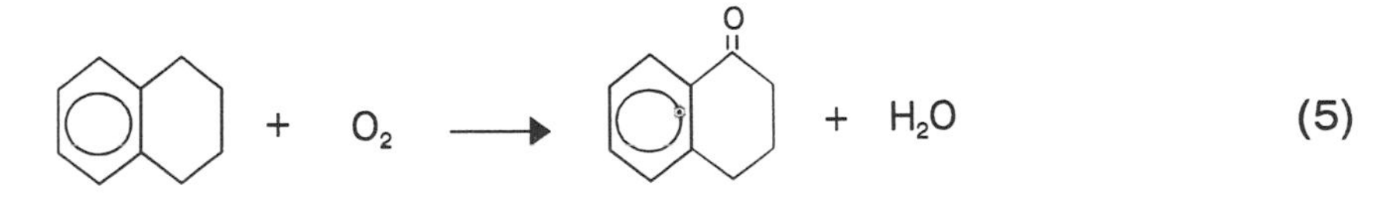

of interest because α-tetralone, its main reaction product, is the starting material for the production of α-naphthol. The side products of the reaction are α-tetralol and the further oxidation products of tetralone and tetralol. Because tetralin is a high-boiling liquid, it can serve as the organic phase, in addition to being the substrate. The homogeneous reaction has been studied in acetic acid and other solvents (*14,15*).

Either manganese, chromium, or nickel chloride can be used in conjunction with N,N,N',N'-tetramethylethylenediamine (TMEDA) and DSS to effect the biphasic autoxidation reaction at 60°C and 1 atm. Triethylphosphite and 2,2,6,6-tetramethyl-3,5-heptanedione are less effective ligands than TMEDA. Nonionic and cationic surfactants are either ineffective or slightly inhibiting.

Of the catalysts studied, manganese gave the highest activity but the lowest tetralone selectivity (60%). Chromium gave the lowest activity (about half of that of manganese) but the highest tetralone selectivity (95%). Nickel gave intermediate activity (about 3/4 of that of manganese) and tetralone selectivity (90%). In each case, the biphasic reaction showed the characteristic features of a free-radical reaction, namely, the presence of an induction period and a sensitivity to free-radical scavengers.

The reaction has an apparent activation energy of 21 kcal/mol, a first-order dependence on the catalyst concentration, a pseudo-zero-order dependence on the tetralin concentration, and a variable oxygen dependence that goes from first to zero order as the oxygen partial pressure is increased above 0.16 atm.

The maximum achievable conversion of the bulk tetralin is about 35% due to the formation of higher oxidation products that inhibit the reaction. A similar product-inhibiting effect has been observed in the homogeneous reaction carried out in acetic acid (*14*); it is believed to arise from product species such as α-naphthol and 1,6-dihydroxynaphthalene.

Notwithstanding the product inhibition effect that limits the tetralin conversion to 35%, the biphasic autoxidation reaction is nevertheless advantageous over the homogeneous reaction for the following reasons. First, the use of noxious and corrosive acetic acid as a reaction medium is avoided. Second, the catalyst complex, being surface-active, can be easily recovered at the interfacial layer and reused. Third, by means of distillation, the unreacted tetralin can be easily separated from the reaction products--including the inhibiting products--and recycled. By contrast, in the homogeneous reaction, a similar recovery and recycle of the catalyst and the unreacted tetralin would be difficult and expensive. The feasibility of recycling the catalyst and the unreacted tetralin in the biphasic technique has, in fact, been confirmed. The recycled runs give essentially the same results as the fresh-catalyst and fresh-tetralin runs.

Comparison with Phase-Transfer, Micellar and Other Biphasic Techniques

The interfacial technique is similar to phase-transfer, micellar and other biphasic techniques (*16-21*) in the use of a biphasic medium, but there are important, fundamental differences. Specifically, the reaction occurs at or near the organic-water interface in the interfacial technique, in the organic phase in the normal phase-transfer technique, in micelles in the micellar technique and in the aqueous phase in the biphasic technique which others have developed recently based on the use of water-soluble ligands (*18-21*). The function of the complexing agent is to draw the catalyst to the interface in the interfacial technique, whereas in the phase-phase technique it is to draw a reactant from one bulk phase to another (usually from aqueous to organic phase). Surface affinity is a desirable property of the complexing agent in the interfacial technique but not in the phase-transfer technique because it would increase resistance to phase transfer. The interfacial reaction is dependent on the interfacial area and should therefore be sensitive to parameters that affect phase dispersion, including stirring, emulsification, and organic-water phase ratio. A phase-transfer reaction would be sensitive to the dispersion parameters only in the mass-transfer limit.

The interfacial technique differs from the micellar technique in that the dispersed phase is much larger and has a much greater capacity for holding the reactant(s). Coalescence of the dispersed phase occurs spontaneously on standing, making phase separation and catalyst recovery easier.

The interfacial technique is intermediate between the micellar and phase-transfer techniques with respect to the interfacial area and the size of the dispersed phase, and it may be viewed as their optimal hybrid. It combines a concentration-enrichment capability with a high substrate-holding capacity. This means that a high reaction rate and a high conversion capacity may be achieved simultaneously.

The difference in the use of surface-active and water-soluble ligands in a biphasic technique produces important differences in the results. With a surface-active ligand, the catalyst complex is segregated in a compact interfacial emulsion layer, where the reaction occurs. Catalyst recovery is relatively easy and is unaffected by the solubilities of the reaction products in either of the bulk phases. With a water-soluble ligand, on the other hand, the catalyst complex is dispersed in the aqueous phase, where the reaction is designed to occur unless the substrate has a low aqueous solubility and cannot be drawn into the aqueous phase by the use of a reverse phase-transfer agent. In the latter event, the reaction will occur at the interface or in the organic phase. In the first case, the catalytic efficiency will be low since only a small fraction of the water-soluble catalyst complex can be found at the interface. In the second case, a phase-transfer agent will be needed to transport the catalyst complex into the organic phase, but this will negate much of the advantage of using a water-soluble ligand for catalyst recovery. The catalyst is recovered in a bulk aqueous phase and the recovery is adversely affected if any of the reaction products is appreciably soluble in the aqueous phase.

The surface-active ligand also makes possible other advantages associated with the interface, namely, a shorter diffusional pathway that lowers the possibility of the reaction being limited by mass transfer, a higher reaction rate on account of the ability of the interface to draw the reacting species into intimate contact, and the possibility of a stereoselectivity control through the directional influence of the interface on the molecular orientations of the reacting species. These advantages are difficult to achieve with the use of a water-soluble ligand. In many ways, the interfacial technique based on the use of surface-active ligands may be viewed as a logical extension and refinement of the biphasic technique which others have developed based on the use of water-soluble ligands.

Conclusion

The feasibility and some of the advantages of a novel organic-water interfacial synthesis technique have been demonstrated on several reactions of commercial and scientific interests. The new technique is based on the use of a surface-active catalyst complex in conjunction with an organic-water mixture and an emulsifier to effect the desired reaction at the organic-water interface. The functions of the reaction medium are decoupled and assigned to a combination of aqueous and organic phases. The new technique is more environmentally benign than conventional liquid-phase syntheses because, first, it avoids the use of toxic or hazardous solvents that may otherwise be needed and, second, it reduces waste generation by making a more efficient use of catalysts, solvents, and reagents. Additionally, the new technique also offers other significant advantages, including the ease of processing and catalyst recovery, a high reactivity, selectivity and reproducibility under mild reaction conditions, and a greater versatility compared to phase-transfer, micellar and other biphasic techniques.

Acknowledgments

Financial support from the U. S. National Science Foundation (Grant No. CTS-9217443-01) and laboratory assistance of Cameron F. Abrams, Veronica M. Godfrey, Brian C. Batts, and Christine J. Yau are gratefully acknowledged.

Literature Cited

1. Hay, A. S. Polymerization by Oxidative Coupling. II. Oxidation of 2,6-Disubstituted Phenols. *J. Polym. Sci.* **1962**, *58*, 581-591.
2. Hay, A. S.; Blanchard, H. S.; Endres, G. F.; Eustance, J. W., Polymerization by Oxidative Coupling. *J. Am. Chem. Soc.* **1959**, *81*, 6335-6336.
3. Hay, A. S. ; Shenian, P.; Gowan, A. L.; Erhardt, P. F.; Haat, W. R.; Theberge, J. E. Phenols, Oxidative Polymerization. *Encyclopedia of Polymer Science and Technology*; Wiley: New York, **1969**, Vol. *10*, pp 92-111.
4. Dautenhahn, P. C.; Lim, P. K. Biphasic Synthesis of Poly(2,6-dimethyl-1,4-phenylene oxide) Using a Surface-Active Coupling Catalyst. *Ind. Eng. Chem. Res.* **1992**, *31*(2), 463-469.
5. Zhong, Y.; Abrams, C. F.; Lim, P. K. Biphasic Synthesis of Poly(2,6-dimethyl-1,4-phenylene oxide) Using a Surface-Active Coupling Catalyst. 2. Process Improvements, Additional Kinetic Results, and Proposed Reaction Mechanism. *Ind. Eng. Chem. Res.* **1995**, *34*(5), 1529-1535.
6. Zhong, Y.; Godfrey, V. M.; Lim, P. K.; Brown, P. A. Biphasic Synthesis of Phenylacetic and Phenylenediacetic Acids by Interfacial Carbonylation of Benzyl Chloride and Dichloro-p-xylene. accepted by *Chem. Eng. Sci.* **1995**.
7. Wilhelm, E.; Battino, R. Thermodynamic Functions of the Solubilities of Gases in Liquids at 25^{0}C. *Chem. Rev.* **1973**, *73*(1), 1-9.
8. Gerrard, W. *Gas Solubilities--Widespread Applications*; Pergamon Press: New York, **1980**.
9. Finkbeiner, H.; Hay, A. S.; Blanchard, H. S.; Endres, G. F., Polymerization by Oxidative Coupling. The Function of Copper in the Oxidation of 2,6-Dimethylphenol. *J. Org. Chem.* **1966**, *31*, 549-555.
10. Tsuchida, E.; Kaneko, M.; Nishide, H. The Kinetics of the Oxidative Polymerization of 2,6-Xylenol with a Copper-Amine Complex. *Makromol. Chem.* **1972**, *151*, 221-234.

11. White, D. M. Reactions of Poly(phenylene Oxide)s with Quinones. *J. Polym. Sci. PC* **1981**, *19*, 1367-1383.
12. Mijs, W. J.; van Lohuizen, O. E.; Bussink, J.; Vollbracht, L., The Catalytic Oxidation of 4-Aryloxyphenols. *Tetrahedron* **1967**, *23*, 2253-2264.
13. Perry, R. J. Making Aramids Using CO and Palladium Catalysis. *Chemtech* **1994**, *24*(2), 18-23.
14. Martan, M.; Manassen, J.; Vofst, D. Oxidation of Tetralin, α-Tetralol and α-Tetralone. *Tetrahedron* **1970**, *26*, 3815-3827.
15. Kamiya, Y.; Ingold, K. U. The Metal-Catalyzed Autoxidation of Tetralin. IV. The Effect of Solvent and Temperature. *Can. J. Chem.* **1964**, *42*, 2424-2433.
16. Starks, C. M.; Liotta, C. L.; Halpern, M. *Phase-Transfer Catalysis: Fundamentals, Applications, and Industrial Perspectives*; Chapman & Hall: New York, **1994**.
17. Fendler, J. H.; Fendler, E. J. Principles of Micellar Catalysis in Aqueous Solutions. *Catalysis in Micellar and Macromolecular Systems*; Academic Press: New York, **1975**; pp 86-103.
18. Cornils, B.; Wiebus, E. Aqueous Catalysts for Organic Reactions. *Chemtech* **1995**, *25*(1), 33-38.
19. Herrmann, W. A.; Kohlpaintner, C. W. Water-Soluble Ligands, Metal Complexes, and Catalysts: Synergism of Homogeneous and Heterogeneous Catalysis. *Angew. Chem. Int. Ed. Engl.* **1993**, *32*, 1524-1544.
20. Kalck, P.; Monteil, F. Use of Water-Soluble Ligands in Homogeneous Catalysis. *Adv. Organomet. Chem.* **1992**, *34*, 219-284.
21. Kuntz, E. G. Homogeneous Catalysis in Water. *Chemtech* **1987**, *17*(9), 570-575.

RECEIVED September 21, 1995

BIOTECHNOLOGY

Chapter 14

Environmentally Efficient Management of Aromatic Compounds

Economic Perspectives of Biocatalytic Conversion of Aromatics to Optically Pure Synthons for the Pharmaceutical Industry

T. Hudlicky[1]

Department of Chemistry, Virginia Polytechnic Institute and State University, Blacksburg, VA 24061

A strategy of economically viable biocatalytic degradation of aromatic compounds is described. Whole-cell fermentation of aromatic compounds by either *Pseudomonas putida* 39D, a blocked mutant, or *Escherichia coli* JM109, a recombinant organism, renders enantiomerically pure arene *cis*-dihydrodiols from the broth. Three topologically similar enzymes recognize specific sites for single ring, fused polycyclic, and biphenyl type aromatics. The diene diol metabolites are excellent synthons for further chemical functionalization. Short synthetic sequences are described that use these synthons in concise preparations of carbohydrates and alkaloids of medicinal importance. The combination of enzymatic degradation and chemical synthesis offers a unique opportunity to shorten routes to important targets in environmentally conscious processes. The review of synthetic accomplishments is concluded with a discussion of future perspectives for reducing chemical pollution at the source.

Aromatic compounds, which may originate from petroleum, coal, plant mass, or industrial outflow, are ubiquitous in the environment. Whether their presence is beneficial or harmful is a function of their inherent properties and their toxicities, as

[1]Department of Chemistry, University of Florida, P.O. Box 117200, Gainesville, FL 32611–7200

0097–6156/96/0626–0180$12.00/0

well as the overall management of their storage, use, and disposal. For example, chlorinated biphenyls, naphthols, and dichlorobenzenes are viewed as contaminants, as are some chlorinated herbicides and pesticides that end up in the soil as a result of various commercial endeavors. The waste stream from chemical industries may contain chlorinated and brominated aromatic compounds, byproducts of manufacturing. The disposal of these compounds is costly, and the development of alternatives to waste-generating components of chemical manufacturing requires long-term commitment to research.

The economical and environmentally-concious utilization of aromatic compounds would accomplish several goals. First, it would lead to recycling of undesirable aromatic waste mass. Second, it would eliminate the need for disposal altogether if the mass could be incorporated into a useful end-product. Third, the conversion of these aromatic compounds by biocatalytic means would likely shorten the synthetic routes and therefore tend to decrease the mass of reagents and solvents used in the overall process. The waste streams of most fermentations, including those that use recombinant microorganisms, are acceptable for discharge into municipal sewer systems, following appropriate sterilization procedures. The cost of disposal of any aromatic compound can easily be compared to the cost of conversion to a target with a defined sale value.

In light of these considerations, this essay seeks to explore the potential of enzymatic conversion of aromatic compounds to optically pure materials of benefit to manufacturing concerns. Shown in Figure 1, in a generalized format for all aromatics, is a two-tier strategy to utilize soil bacteria, as blocked mutants or recombinant organisms, to degrade aromatics to *cis*-diols of type **1**, useful as intermediates in asymmetric synthesis. The judicious use of known soil organisms and their degradative pathways, coupled with clever synthetic design, leads to efficient preparation of valuable synthons with immediate relevance in the preparation of medicinal agents and other useful compounds.

There are three major classes of aromatic compounds, and three types of closely related enzymes evolved to degrade them for energy and carbon utilization. Single-ring aromatics are oxidized to *cis*-diols **1** (portrayed by the solid-line structure in Figure 1) by toluene dioxygenase, a multi-component oxidoreductase. The enzyme introduces the diol unit with remarkable specificity, almost always in the 2,3 position with respect to group X, and the enzyme tolerates a large number of functional groups in this position. Both naphthalene dioxygenase, specific for fused

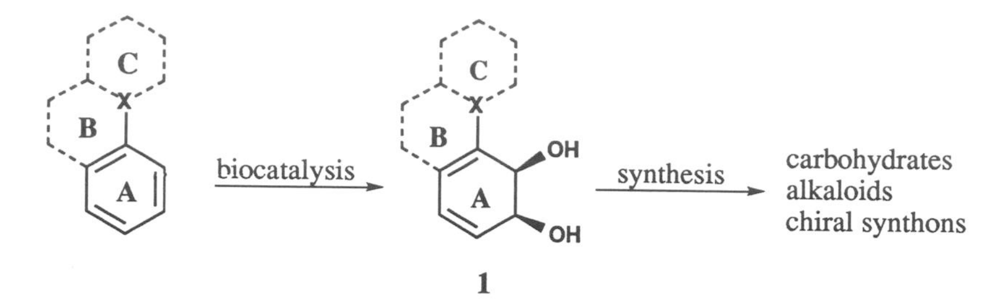

Single-ring aromatics. Ring A only. X = H, Cl, Br, I, F, alkyl, CN, etc.
Naphthalenes. Rings A and B. X = CH
Phenanthrenes. Rings A, B, and C. X = C
Biphenyls. Rings A and C. X = C

Figure 1. Two-tier Strategy for Conversion of Halogenated Aromatic Compounds.

polycyclic aromatics, and biphenyl dioxegenase also operate with regio-, stereo- and enantiospecificity. The representation in Figure 1 accounts for all known cases except those derived from benzoic acid and substituted benzoic acids, where the regiochemistry is sometimes 1,2- with respect to the directing group X. Some heterocyclic aromatics have also yielded the corresponding diol metabolites. A brief history of this field, presentation of recent accomplishments, and an indication of the power and the broad applicability of this methodology in the overall context of pollution prevention at the source are the focus of this manuscript.

Results

In the late 1960s David Gibson elucidated the oxidative degradation pathway of aromatic compounds by *Pseudomonas putida*, a common soil organism (*1-2*). A blocked mutant lacking the capability of synthesizing catechol dehydrogenase was developed by his group, followed later by a recombinant organism *E. coli* JM109, which required no induction by aromatic compounds (*3-5*). This chemically unprecedented transformation was slow to gain recognition in the synthetic community; the first application was not until 1987 when Ley published the conversion of the *cis*-diol derived from benzene to racemic pinitol **3**, Figure 2 (*6*). Then Imperial Chemical Industries (U. K.) used diol **2** (*7*) in polyphenylene manufacturing, and in 1988 was the disclosure of an efficient preparation of enone **5**, an intermediate for prostaglandin synthesis, from toluene (*8*). The first example of chemical conversion of an aromatic ring to a *cis*-dihydrodiol of type **2** was not reported until 1995, when Motherwell described a cyclitol synthesis utilizing a photochemical oxidation of benzene in the presence of OsO_4 (*9*).

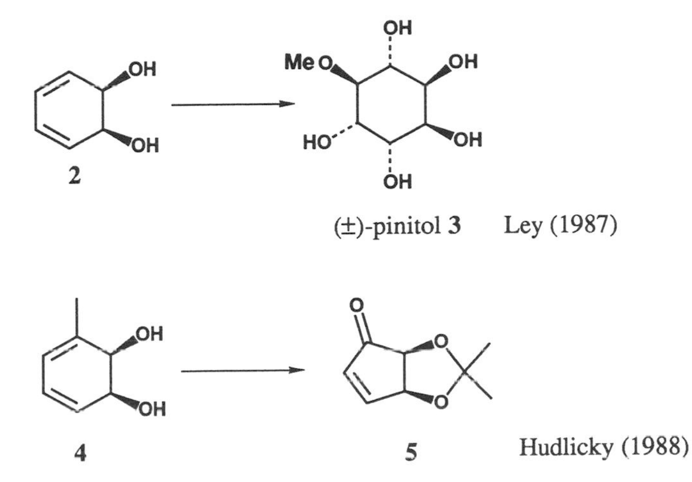

Figure 2. Early Examples of the Use of Arene cis-Dihydrodiols in Synthesis.

The use of metabolites derived from aromatics has undergone explosive growth, and by 1995, the time of the latest of several reviews (*10-14*), there were several hundred publications reporting isolation and structure elucidation of new metabolites as well as the use of these compounds in enantioselective syntheses. To date some 200 *cis*-diols derived from benzene derivatives, fused polycyclic aromatics, biphenyls, and heteroaromatic compounds are known (*15-16*). The enzymes, specific for each of these classes of compounds, display remarkable specificity for the regiochemistry of these oxidations and the absolute stereochemistry, while tolerating a variety of substituents. In Figure 3, the more typical of the diol structures, along with the corresponding enzyme, demonstrate the structural variations possible in the substrate.

Several groups in the United Kingdom and two in the United States began to exploit the asymmetric features of these diverse and potentially valuable synthons in synthesis. A look at a general design for entire classes of compounds from the simple, optically pure materials provided by the enzymes make the advantages of a chemoenzymatic approach to the preparation of complex molecules clear.

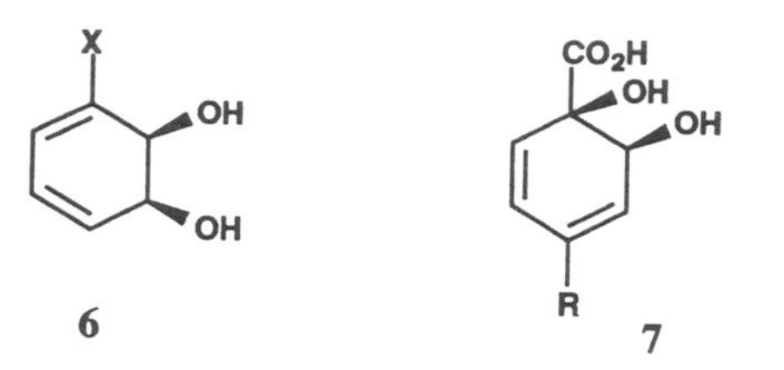

toluene dioxygenase

R = H, CH_3

X = H, Cl, Br, alkyl, aryl, etc.

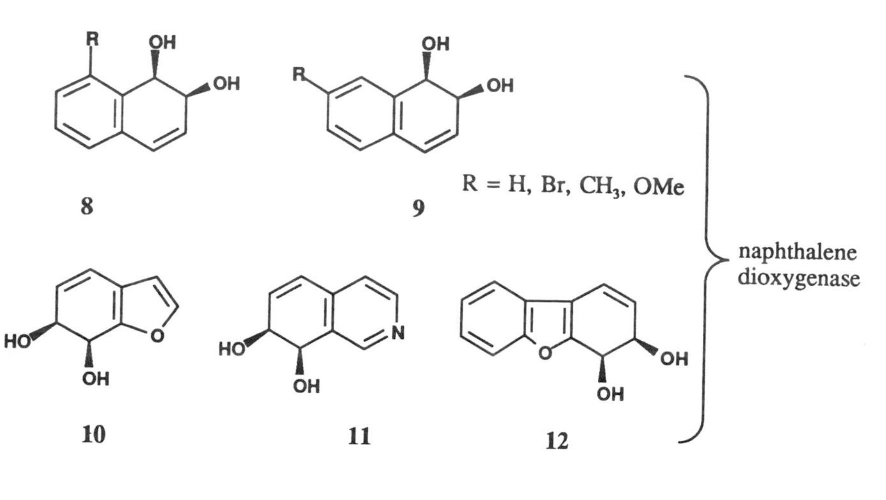

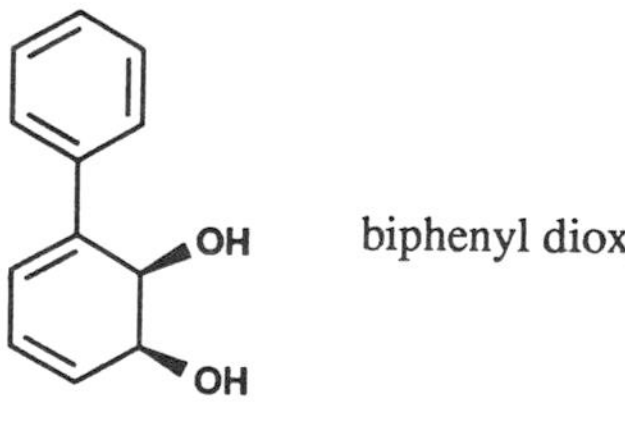

biphenyl dioxygenase

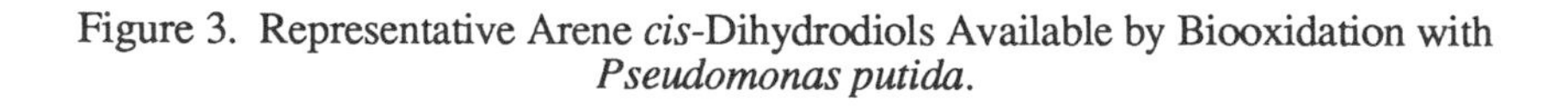

Figure 3. Representative Arene *cis*-Dihydrodiols Available by Biooxidation with *Pseudomonas putida*.

Design Elements

There are a number of stereocontrolled transformations possible from the multiple functionalities of diols of type **6** (See Figure 4). The *cis*-diol unit provides a unique opportunity to direct the next functionalization either to the same face (when R = H) or the opposite face if the diol is protected with a bulky group such as an acetonide. In this way the diastereoselectivity of the next center introduced is established.

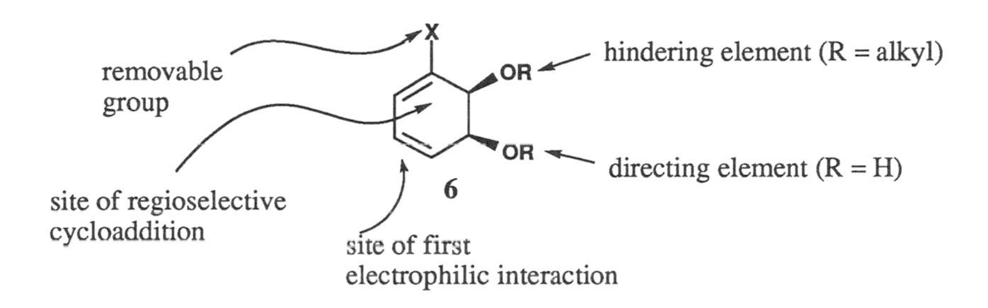

Figure 4. Design Features of Diene-diols **6** for Controlled Synthetic Operations.

The substituent X most important to the reactivity and symmetry of the diols is by far the halogen. The diene unit, when polarized by its presence, reacts readily in many regioselective cycloadditions. Additionally, the olefin not bearing the halogen atom (in diols where X ≠ Cl, Br, I) allows the use of standard electrophilic reagents. Because the halogen can be easily removed reductively following the funtionalization of the more electron-rich olefin, it allows for the enantiodivergent design of a synthesis of both target isomers from a single enantiomer of the starting diol. This feature, extremely important in the manufacture of potential pharmaceuticals, is best illustrated by the enantiodivergent synthesis of (+)- and (–)-pinitols and (+)- and (–)-*chiro*-inositols, as shown in Figure 5 (*17*).

The recognition of symmetry relationships in the target compounds, i.e., that the RO and HO groups can be transposed in a 1,2-fashion, allows the decision of introducing first into the starting diene **14** (X = Cl or Br) either the *cis*-hydrin or the *trans*-hydrin unit, as in the cases of **17** and **15**, respectively. The reagents and conditions of the synthesis of either path are identical, only the order changes. Only stereochemistry and regiochemistry are affected by the choice and the ordering of reagents. The necessary symmetry switch between the (+)- and the (–)- space is

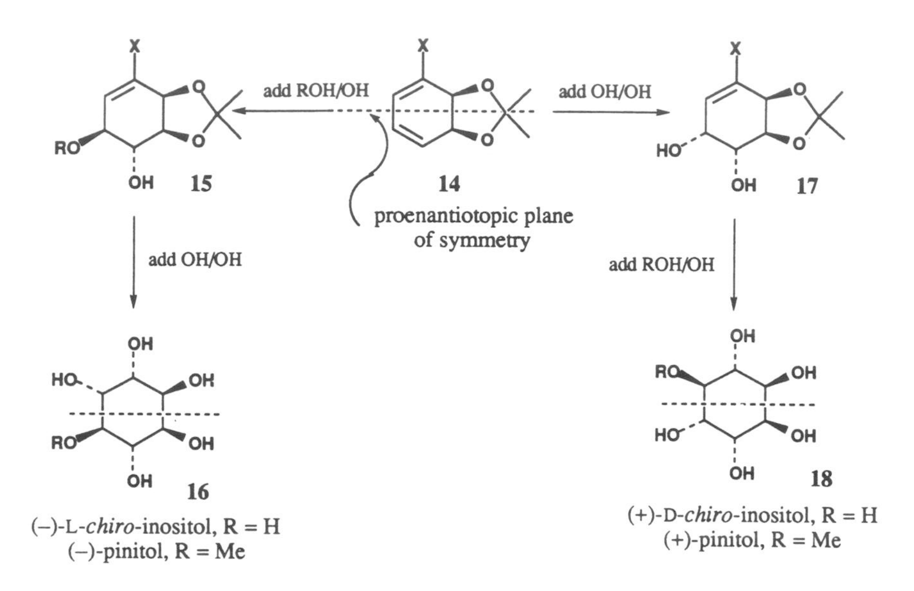

(–)-L-*chiro*-inositol, R = H
(–)-pinitol, R = Me

(+)-D-*chiro*-inositol, R = H
(+)-pinitol, R = Me

Figure 5. Enantiodivergent Synthesis of Inositols.

made in the first sequence. The dashed line in Figure 5 can be described as a "plane of proenantiotopic symmetry" because an imaginary removal of the halogen atom from the intermediate would symmetrize the compound. This feature is useful in the design of the enantiomeric switch during the forward execution of the synthesis. (For a more detailed discussion and definition of these terms see ref. *11*, *17*, and *18*).

With this background, an exhaustive program of rational design for carbohydrates and alkaloids was initiated. In the planning stages of the design for carbohydrates, the functionalization of diols **6** and its acetonide **14**, and the selective cleavage and recyclization of these compounds was viewed as an important feature. Portrayed in Figure 6 is the rationale for the approach to all major classes of carbohydrates. The controlled introduction of hydroxyls, amino groups, or other heteroatomic functionalities into either diols **6** or acetonides **14** lead to inositols **19** or aminocyclitols **20**. The stereochemical features are controlled by either directing or hindering effects of the previously installed centers, as portrayed in Figure 4. The functionalization of the diene of acetonide **14** was demonstrated in earlier detailed studies and syntheses of several cyclitols (*18-21*). These syntheses take advantage of

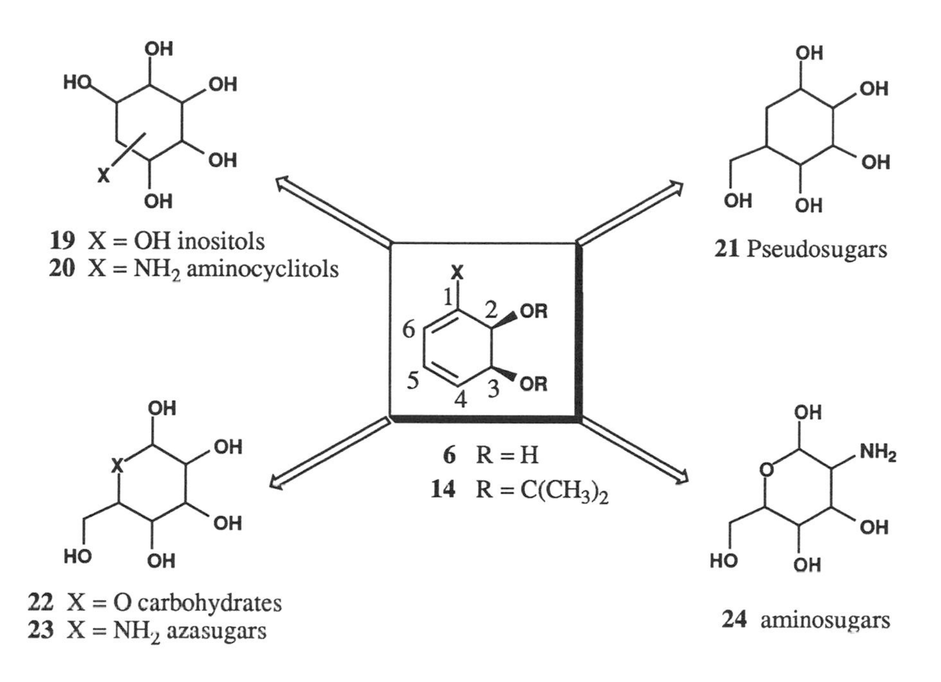

Figure 6. General Design for Carbohydrate Synthesis.

either α- or the β-epoxides installed at the electron-rich olefinic site and the subsequent trans-diaxial opening of these epoxides with heteratom nucleophiles, as shown below.

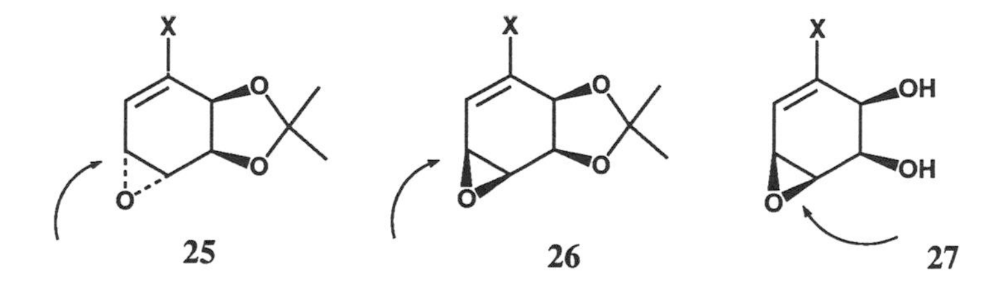

The introduction of additional carbon functionality at C-1 is possible via organometallic coupling, and, when followed by further oxygenation of the periphery of the diene–diol, this approach yields pseudopyranoses of type **21**, as has been recently demonstrated (*22*). To synthesize pyranoses, as well as lower carbohydrates, oxidative cleavage of a selected edge of the cyclitol or inositol is required. This is easily accomplished by functionalizing the C4-C5 olefin either as a diol (for pyranose approach) or as an aminohydrin (for the azasugar approach). Following precise placement of these substituents, the halogenated olefin side, or a suitably freed diol, is oxidatively cleaved releasing latent aldehyde(s) and/or carboxylate(s). Either of these groups can then be directed to receive whatever nucleophile is available and disposed for a reclosure to a pyranose system. Similarly, a cyclization of an unprotected alcohol onto one of these termini in a substrate that contains a protected nitrogen substituent leads to pyranoses with an *exo*-situated nitrogen in either the 2-, 3-, or 4-position. Figure 7 is a summary of recently achieved results in this area and demonstrates the generality of execution of any topological and diasteromeric combination in the carbohydrate domain (*23-29*). Thus both the topology of the different types of carbohydrates and the topography of individual compounds is accounted for in this general design. These examples accurately portray the broad applicability of this methodology especially to access unnatural derivatives of carbohydrates, as such compounds are normally available only by tedious and lengthy procedures from natural sugars.

In the alkaloid domain, several hydroxylated alkaloids were chosen to demonstrate that arene *cis*-dihydrodiols can serve as synthons in heterocyclic synthesis. The first venture into this area was the enantiodivergent synthesis of pyrolizidine triols from erythruronolactone **30**, as shown in Figure 8 (*30-32*). The same principles of enantiodivergent design of the pinitols were applied here to yield the two enantiomeric azido dienes **40** and **41**, which were then transformed to the alkaloids by previously documented azide–diene cycloaddition followed by vinylaziridine–pyrroline rearrangement (*33*).

Lycoricidine **46** was approached next by recognizing that the C ring of this alkaloid is an aminoconduritol easily attainable via a nitrosyl Diels–Alder reaction. The final carbon–carbon bond formation between the aryl unit and the aminoconduritol was made by means of the Heck reaction, as illustrated in Figure 9 (*34-35*).

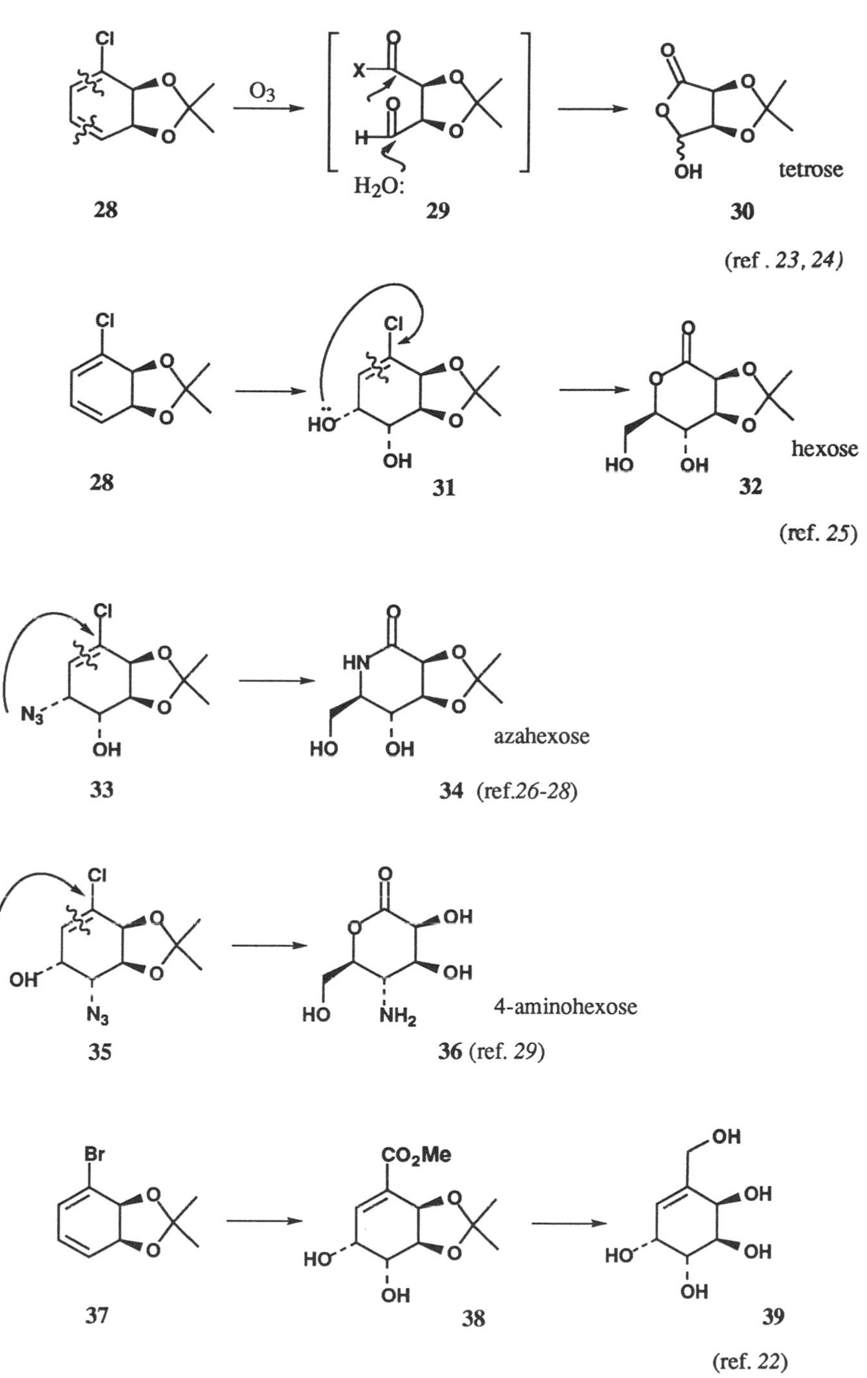

Figure 7. Recent Examples of Sugar Synthesis.

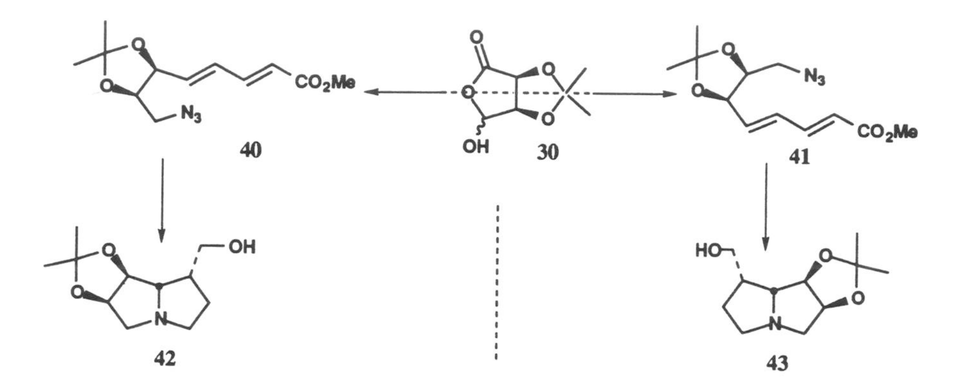

Figure 8. Enantiodivergent Trihydroxyheliotridane Synthesis.

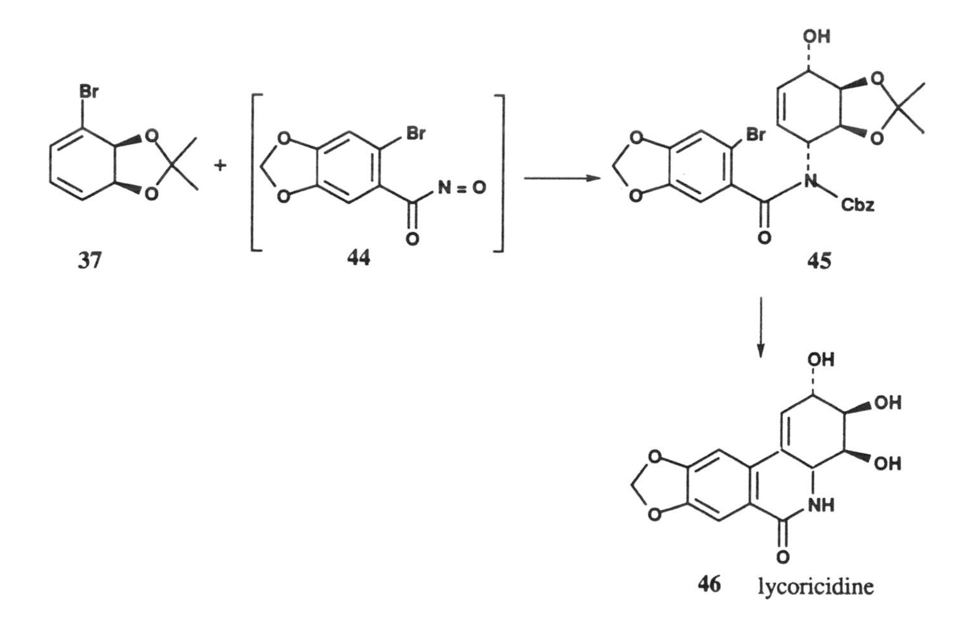

Figure 9. Asymmetric Synthesis of Lycoricidine.

A more sophisticated target, for which an asymmetric preparation had not existed at the onset of our study, was pancratistatin **52**, a promising antitumor agent in short supply from natural sources. This target was approached by a completely new strategy, a part of a systematic approach to alkaloids and oligosaccharides from vinylaziridine **47** and vinylepoxide **48**, both derived from bromobenzene as shown in Figure 10 (X = Br, H).

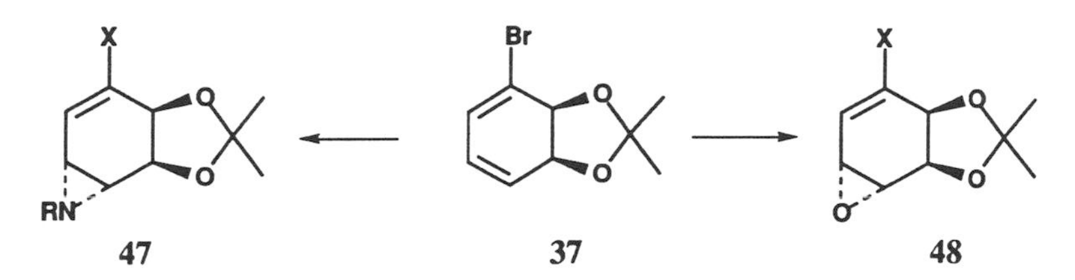

Figure 10. Alkaloid and Oligosaccharide Synthons.

The chemistry of these systems was investigated in detail especially with regard to the regiochemistry of their interactions with carbon and heteratom nucleophiles (*36*). A relatively short synthesis of (+)-pancratistatin ensued, featuring the addition of a mixed cuprate to aziridine **47**, Figure 11 (*37*). Following the attainment of amide **50**

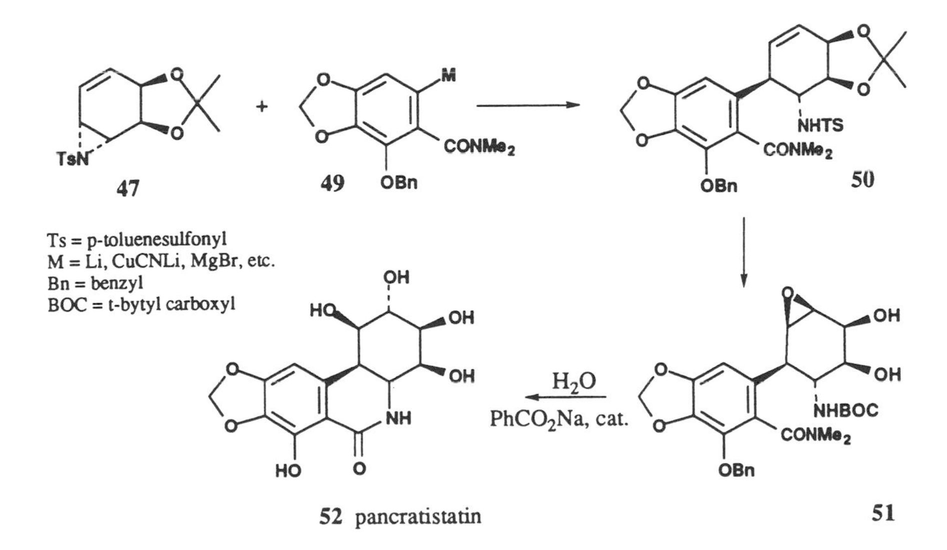

Figure 11. First Asymmetric Synthesis of Pancratistatin.

and some adjustments in functionality leading to epoxy diol **51**, a remarkable transformation was observed on attempted hydrolysis of the epoxide. Under the conditions previously used to hydrolyze epoxides with concomitant deprotection of acetonides (*38*), the deprotection of BOC group, transamidation, and, amazingly, debenzylation took place to furnish (+)-pancratistatin in a total of thirteen steps from bromobenzene. Further refinements in this synthesis will no doubt lead to an enhanced supply of this promising alkaloid.

The latest venture addresses the synthesis of both natural and unnatural oligosaccharides. To this end model studies were directed at developing methods for the coupling of epoxides **48** with heteroatom and carbon nucleophiles. The epoxides of type **48** were subsequently converted to the nor-disaccharide analogs **53** and **54**, compounds serving as models for an approach to pseudo-oligosaccharadies that promise an exciting array of biological activities because of their resemblance to natural sugar conjugates with metabolic functions in insulin mediation and glycolysis respectively (*39-40*).

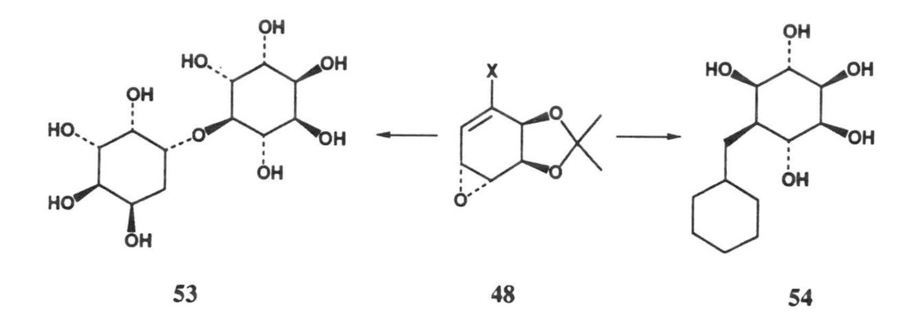

Conclusion

The above examples demonstrate the enormous value of the arene *cis*-diols in asymmetric synthesis and also underscore the immense potential of biocatalysis and the use of enzymes to achieve selected organic transformations. The primary reasons to engage in biocatalytic preparation of organic compounds are, of course, the advantages offered by the aqueous medium of the reaction and the potential of introducing asymmetry through enzyme-catalyzed transformations. Additionally, properly planned biocatalytic sequences take advantage of the best possible combination of an enzymatic processes followed by synthetic transformations. The result will be expressed ultimately in brief synthetic schemes and in the overall efficiency of a particular process. A common-sense analysis of the value of any synthetic venture with respect to the amount of byproduct mass will clearly indicate that the shorter the process or the fewer the reagents or their mass, the less the waste mass produced. Thus it seems senseless in an environment-conscious age to focus on designing esoteric reaction sequences involving costly or toxic reagents. The era of 30-or-more-step syntheses of target compounds seems to be drawing to a close. The responsible preparation of organic compounds by either degradation of waste products or by *de novo* enzymatic synthesis of targets from renewable resources seems to be the way of the future.

Aromatic compounds will always be used in the chemical and pharmaceutical industry. Whether they originate from petroleum products or whether they are manufactured by recombinant organisms, as in the research program of Frost (*41*) for example, is less important an issue than the consideration of their effect on the environment. What does matter is the maintenance of balance and overall management of chemical entities to benefit the end-use of products.

Biocatalysis is a new discipline that is marvelously expressed in environmentally-conscious chemical manufacturing. It is somewhat surprising that its acceptance in the U.S. has been rather slow, in comparison with Europe, where every effort is directed toward promulgation of "green" methods of synthesis. In the U.S. however, the synthetic community is tenaciously holding on to old methods of synthesis and has not, for the most part, made the logical and evolutionary adjustments toward new and obviously better means to achieving its goal, which has always been and always will be the preparation of organic compounds in useful quantity. A pertinant diagram is shown in Figure 12. It shows the various ways in which existing or new aromatic compounds can be either recycled or prepared by well-thought-out processes.

Relevance to the Prevention of Pollution at the Source

The diagram in Figure 12 is especially significant in terms of the role of bicatalysis in enviromentally responsible manufacturing. The efficiency of a synthetic scheme is usually dictated by the overall number of steps, the nature, mass, and cost of reagents, the volume of solvents, the need for chromatographic separation, the energy required to either heat or cool the reaction medium, and other

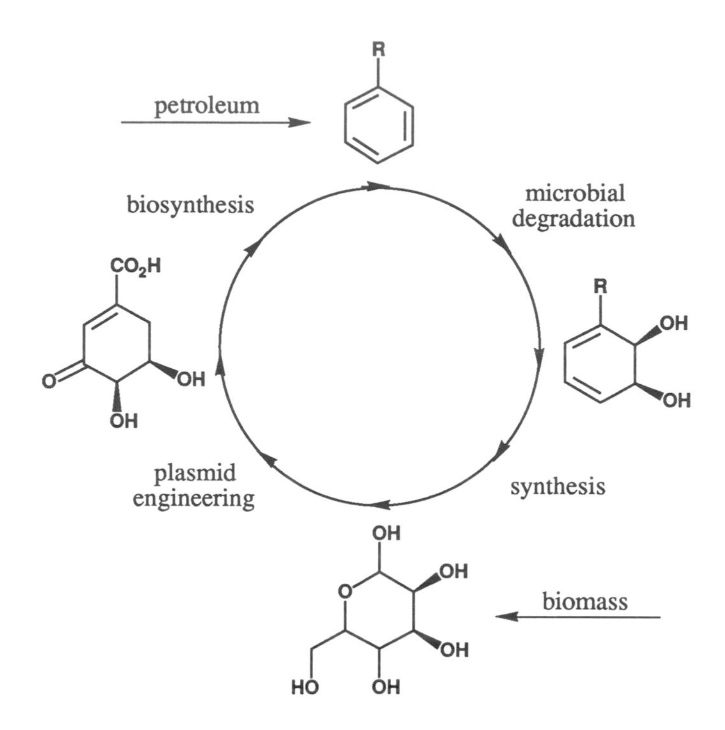

Figure 12. Recycling Scheme for Aromatic Compounds.

factors. The enzymatic preparation of chiral compounds can, in a single reaction step, produce a complex intermediate with absolute stereochemistry and in an aqueous medium. The organism that is used to accomplish this task can be grown on available renewable resource such as glucose, thus the cost of the fermentation is probably lower than the equivalent chemical process. The high chemical complexity of the product obtained in the biocatalytic step lends itself to further chemical manipulations to the final target. Overall, such routes will most likely be significantly shorter (less mass input), contain steps that are performed in water (less solvent volume), use fewer reagents (less waste mass), and provide enantiomerically pure compounds. All of the syntheses described in this paper have been attained in approximately half of the steps of those preparations that used traditional chemical routes. In the scenario described for the degradation of aromatic compounds the major benefit of the biocatalytic handling is the conversion of (perceptibly) undesired materials to intermediates with a defined value, thereby saving resources in both waste disposal and redundant manufacturing. Most significant is the connection between enzymatic degradation of arenes and enzymatic production of arenes. The exploitation of the two diverse pathways can take place in two directions. The examples discussed above represent first-generation academic efforts, and these can be further streamlined in process development in those cases that warrant end use.

The efficiency that is achieved by the combination of biological and chemical methods with well thought-out design of the overall synthetic plan is self-evident. Every effort should be made to restructure the manufacturing of chemical intermediates for the pharmaceutical industry to include chemoenzymatic procedures. In the future there will no doubt be further examples of exploitation of this new area of organic synthesis as well as new interdisciplinary mergers in the fields of chemistry and biology.

Acknowledgments

The research presented in this manuscript would not have been possible without the excellent effort of the many coworkers whose names appear in the citations. Financial support for the biocatalysis program was provided by Jeffress Trust Fund; Genencor International, Inc.; Mallinckrodt Specialty Chemicals, Inc.; TDC Research, Inc.; TDC Research Foundation; and the National Science Foundation (Environmentally Benign Synthesis Initiative with the EPA).

Literature Cited

1. Gibson, D.T.; Hensley, M.; Yoshioka, H.; Mabry, T.J. *Biochemistry* **1970**, *9*, 1626.
2. Gibson, D.T.; Subramanian, V. In *Microbial Degradation of Organic Compounds;* Gibson, D.T., Ed.; Microbiology Series, Vol. 13, Marcel Dekker: New York, 1984; Chapter 7.
3. Gibson, D.T.; Zylstra, G.J.; Chauhan, S. In *Pseudomonas: Tiotransformations, Pathenogenesis, and Evolving Biuochemistry;* Silver, S.; Chakrabarty, A.M.; Iglewski, B.; Kaplan, S., Eds.; Am. Soc. for Microbiol., 1990; Chapter 13, p. 121
4. Zylstra, G.J.; Gibson, D.T. In *Genetic Engineering*; Setlow, J.A. Ed.; Pergamon: New York, 1991; Vol 13, p 183

5. Zylstra, G.J.; Gibson, D.T. *J. Biol. Chem.* **1989**, *264*, 14940.
6. Ley, S.V.; Sternfeld, F.; Taylor, S. *Tetrahedron Lett.* **1987**, *28*, 225.
7. Ballard, D.G.H.; Courtis, A.; Shirley, I.N.; Taylor, S.C. *Macromolecules* **1988**, *21*, 294.
8. Hudlicky, T.; Luna, H.; Barbieri, G.; Kwart, L.D. *J. Am. Chem. Soc.* **1988**, *110*, 4735.
9. Motherwell, W. B.; Williams, A. S. *Angew. Chem. Int. Ed. Engl.* **1995**, *34*, 2031.
10. *Chemical and Engineering News,* September 5, 1994.
11. Hudlicky, T.; Reed, J.W. In *Advances in Asymmetric Synthesis;* Hassner, A., Ed.; JAI Press: Greenwich, CT, **1995**, Vol. 1, p. 271.
12. Brown, S.M.; Hudlicky, T. In *Organic Synthesis: Theory and Practice;* Hudlicky, T., Ed.; JAI Press: Greenwich, CT, **1993**; Vol. 2, p. 113.
13. Carless, H.A.H. *Tetrahedron: Asymm.* **1992**, *3*, 795.
14. Widdowson, D.A.; Ribbons, D.W.; Thomas, S.D. *Janssen Chimica Acta* **1990**, *8* (3), 3.
15. Stabile, Michele R., Ph.D. thesis, Virginia Polytechnic Institute and State University, 1995.
16. McMordie, R.A.S. Ph.D. thesis, School of Chemistry, The Queen's University of Belfast, Belfast BT9 5AG, U.K., 1991.
17. Hudlicky, T.; Rulin, F.; Tsunoda, T.; Price, J.D. *J. Am. Chem. Soc.* **1990**, *112*, 9439.
18. Hudlicky, T.; Luna, H.; Olivo, H.F.; Andersen, C.; Nugent, T.; Price, J.D. *J. Chem. Soc. Perkin I* **1991**, 2907.
19. Hudlicky, T.; Olivo, H.F. *Tetrahedron Lett.,* **1991**, *32*, 6077.
20. Hudlicky, T.; Price, J.D.; Olivo, H.F. *Synlett* **1991**, 645.
21. Hudlicky, T.; Rulin, Olivo, H.F.; Andersen, C.; Nugent, T.; Price, J.D. *Isr. J. Chem.* **1991**, *31*, 229.
22. Hudlicky, T.; Entwistle, D.A. *Tetrahedron Lett.* **1995**, *36*, 2591.
23. Hudlicky, T.; Mandel, M.; Kwart, L.D.; Whited, G.M. Collect. *Czech. Chem. Commun.* **1993**, *58*, 2517.
24. Hudlicky, T.; Luna, H.; Price, J.D.; Rulin, F. *J. Org. Chem.* **1990**, *55*, 4683.
25. Hudlicky, T.; Mandel, M.; Rouden, J.; Lee, R.S.; Bachmann, B.; Dudding, T.; Yost, K.; Merola, J.S. *J. Chem. Soc. Perkin Trans. 1* **1994**, 1553.
26. Hudlicky, T.; Rouden, J.; Luna, H. *J. Org. Chem.* **1993**, *58*, 985.

27. Hudlicky, T.; Rouden, J. *J. Chem. Soc. Perkin 1* **1993**, 1095.
28. Hudlicky, T.; Rouden, J.; Luna, H.; Allen, S. *J. Am. Chem. Soc.* **1994**, *116*, 5099.
29. Hudlicky, T.; Pitzer, K. *Synlett* **1995**, 803.
30. Hudlicky, T.; Seoane, G.; Seoane, A.; Frazier, J.O.; Kwart, L.D.; Tiedje, M.H.; Beal, C. *J. Am. Chem. Soc.* **1986**, *108*, 3755-3762.
31. Hudlicky, T.; Sinai-Zingde, G.; Seoane, G. *Synth. Commun.* **1987**, *17*, 1155.
32. Hudlicky, T.; Seoane, G.; Lovelace, T.C. *J. Org. Chem.* **1988**, *53*, 2094.
33. Hudlicky, T.; Seoane, G.; Price, J.D.; Gadamasetti, K. *Synlett* **1990**, 433.
34. Hudlicky, T.; Olivo, H.F. *J. Am. Chem. Soc.* **1992**, *114*, 9694.
35. Hudlicky, T.; Olivo, H.F.; McKibben, B. *J. Am. Chem. Soc.* **1994**, *116*, 5108.
36. Hudlicky, T.; Tian, X.; Konigsberger, K.; Rouden, J. *J. Org. Chem.* **1994**, *59*, 4037.
37. Hudlicky, T.; Tian, X.; Konigsberger, K. *J. Am. Chem. Soc.* **1995**, *117*, 3543.
38. Hudlicky, T.; Thorpe, A.J. *Synlett* **1994**, 899.
39. Reddy, K.K.; Falck, J.R.; Capdevila, J. *Tetrahedron Lett.*, **1993**, *34*, 7869.
40. Ley, S.V.; Yeung, L.L. *Synlett*, **1992**, 997.
41. Dell, K.A.; Frost, J.W. *J. Am. Chem. Soc.* **1995**, *115*, 11581.

RECEIVED December 29, 1995

Chapter 15

Environmentally Benign Production of Commodity Chemicals Through Biotechnology

Recent Progress and Future Potential

Leland C. Webster[1], Paul T. Anastas[2], and Tracy C. Williamson[2]

[1]15 Orkney Road, Brookline, MA 02146
[2]Office of Pollution Prevention and Toxics, U.S. Environmental Protection Agency, Mail Code 7406, 401 M Street, S.W., Washington, DC 20460

The United States chemical industry has raised the standard of living in America and has contributed enormously to the country's economic vitality in the twentieth century, but this success has come at a price to the environment. Plentiful and cheap, petroleum has been the dominant feedstock for chemical manufacturing since World War II. Biotechnology has been increasingly investigated as an alternative and has the potential to impact the chemical industry substantially. From its use of renewable, non-toxic biomass in place of petroleum, to the fermentation of genetically engineered microorganisms and novel processing techniques, the development of biotechnology methods for chemical manufacture is quietly taking place. This chapter will discuss some of these emerging technologies including the conversion of cellulose to glucose, the use of microbes as biocatalysts, metabolic pathway engineering, aromatic chemical pathways, bioprocessing, and the use of catalytic antibodies.

One hundred and fifty years ago, the source of most industrial organic chemicals was biomass (i.e., plant matter), with a lesser amount being derived from animal matter *(1)*. After this time, the use of coal as a chemical feedstock surged, and the use of petroleum followed quickly with the discovery of an inexpensive method for the extraction of oil from underground sources. Shortly after World War II, petroleum had become the dominant chemicals feedstock, and today, over 95% of organic chemicals are derived from petroleum *(1)*.

The abundance, affordability and versatility of petroleum permitted the remarkable growth of the chemical industry and the introduction of entirely new chemicals and derivative products. It was not until the late 1960's and 1970's that some of the negative environmental effects associated with chemical manufacturing

0097–6156/96/0626–0198$12.00/0

and processing were widely acknowledged. Since then, the US petrochemical industry has expanded in spite of its significant role in pollution generation, and will likely continue to expand, given its important role in the economy. In addition to the patent disadvantage of heavy US dependence on foreign oil for chemicals (not to mention fuels), the pollution issue remains a major concern.

Biotechnology

The term "biotechnology" is often used synonymously with "biopharmaceutical", given the high profile of the human therapeutics industry. However, biotechnology also encompasses bioremediation, which involves the use of microbes that are genetically engineered to degrade toxic chemical wastes in clean-up operations. While biotechnology is most often associated with the use of recombinant DNA, it is more broadly defined as the engineering and use of living organisms to benefit mankind. A little-known application of biotechnology is the development of methods to produce commodity chemicals, i.e., chemicals manufactured on the same scale as petrochemicals. This endeavor is faced with obstacles on many levels, but recent technological advances suggest an exciting future for an industry that promises to contribute to the generation of environmentally more benign organic chemicals. This chapter focuses on how biotechnology is being brought to bear in this endeavor.

Producing Chemicals Through Biotechnology: Essential Components and Recent Advances

From an ideal perspective, the application of biotechnology to chemicals production begins with biomass as a feedstock instead of petroleum. Glucose and other sugars derived from biomass serve as the chemical starting material in the fermentation of microbes that have been metabolically engineered, or otherwise selected to produce a desired chemical substance. Biochemical engineering techniques are then used to separate the chemical from the fermentation stream. For specific, limited chemical transformation steps, purified enzymes may take the place of microbes as biocatalysts.

Biomass. There are two predominant types of biomass: starch and lignocellulosics. Corn, wheat, sorghum, and potato are representative of the starch class, whereas agricultural wastes (such as corn cobs and stovers, wheat straw, etc.), forestry wastes, and dedicated woody and herbaceous crops comprise the bulk of available and potential lignocellulosics. There is a general consensus that current and future supplies of biomass will not be a limiting factor in the production of organic chemicals *(2)*.

Both starch crops and lignocellulosics contain polymers of sugars that can be broken down into monomers and used in fermentation. The federally-subsidized production of fuel ethanol from corn is an example of bioconversion that takes advantage of well-established wet and dry milling techniques after which the starch is enzymatically converted to glucose for yeast fermentation.

Lignocellulose. Most of the earth's biomass is lignocellulosic, the most abundant component being cellulose. In fact, cellulose is the most abundant organic compound in the biosphere *(3)*, and its exploitation as an inexpensive chemical feedstock is the key to the future success of biomass conversion. Like starch, cellulose can be processed to yield glucose, but with much greater difficulty, for several reasons. First, much of the cellulose is crystalline and therefore resistant to hydrolysis. Second, cellulose has beta-1,4 chemical linkages joining the glucose monomers, which are more resistant to hydrolysis than the alpha-1,4 linkages in starch. Third, cellulose is tightly associated with hemicellulose and is also covalently bound to lignin, a phenyl-propene polymer. This tight association with hemicellulose and lignin impedes cellulose degradation and therefore complicates processing.

Pretreatment of Lignocellulose. To make lignocellulose more accessible to processing, a variety of "explosion" techniques can be used to disrupt its structure *(4)*. In this process, the lignocellulosic material is subjected to highly pressurized steam which is then suddenly released so that ambient pressure is rapidly attained. Other strategies include dilute acid treatment, organosolv techniques, and supercritical extraction *(4)*. Recently, workers at the National Renewable Energy Laboratory (NREL), as part of the United States Department of Energy's Alternative Feedstocks Program (AFP), have developed methods to effectively separate, at efficiencies approaching 100%, the three components of lignocellulose. This so-called clean fractionation is an exciting development, because the resulting purified cellulose component is much more efficiently converted to glucose than is lignocellulosic material that has undergone pretreatment alone. In addition, the purified lignin component itself can be used to produce a variety of chemicals by more traditional chemical methods. A disadvantage is that as a result of the solvents used, one component, the purified hemicellulose, has proven intractable to further manipulation, which means that the lignocellulosic material is not being used to its maximum potential. This problem is probably not insurmountable; in any event, NREL's clean fractionation, already demonstrated on batches of 100 grams, is currently being scaled up *(5)*.

Conversion of Cellulose to Glucose: the Cellulases. A number of microorganisms have been shown to possess cellulase activity, i.e., the ability to break cellulose down into glucose. It has been suggested that the organisms characterized thus far represent just the tip of the iceberg in terms of diversity *(6)*.

Study of the most familiar cellulase-producing microorganism dates back to the 1940's. The fungus *Trichoderma reesei* has been shown to produce a number of enzymes that possess cellulase activity, including endoglucanase, cellobiohydrolase, and beta-glucosidase, which act synergistically *(7)*. Endoglucanase cleaves cellulose at internal glycosidic bonds, and cellobiohydrolase hydrolyzes cellulose from chain ends to yield dimers of glucose, which beta-glucosidase then cleaves into glucose. Although one of the long-standing problems with the *Trichoderma reesei* cellulase has been its inhibition by glucose, novel mutant selection assays have been developed to overcome this limitation, and efforts at strain improvement continue *(8)*. Several structure determinations of cellulase components have been made recently, the most recent of which is for the cellobiohydrolase I catalytic core from *Trichoderma reesei (9)*. The elucidation of cellulase structure-function relationships could lead to protein engineering advances aimed at increasing the effectiveness of cellulases.

More recently, cellulase-producing bacteria have been intensively studied. A thermally stable beta-glucosidase has been cloned from *Microbispora bispora (10)* that retains most of its activity even after 48 hours at 60°C, whereas the *Trichoderma reesei* beta-glucosidase is completely inactivated after 10 minutes at 60°C *(11)*. Furthermore, not only is the *Microbispora bispora* enzyme resistant to end product inhibition, it is actually activated by a range of glucose concentrations.

Another thermophilic bacterium, *Clostridium thermocellum*, possesses a distinct cellulase activity that requires the presence of Ca^{2+} and a reducing agent, and is more active against crystalline cellulose than amorphous cellulose *(12)*. This particular cellulase exists as a cellulosome, an enormous complex that contains at least 15 different proteins *(13)*, all of which have been cloned and sequenced *(14)*. This molecular cloning has enabled researchers to functionally dissect the cellulosome, revealing a tethering mechanism in which the large CelL subunit (M_r = 250,000), itself lacking enzymatic activity, attaches to the cellulose, which then acts as a scaffold to which the enzymatic components of the cellulosome attach *(15)*. The cloning and sequencing of CelS, the gene for the most abundant enzymatic component secreted by *Clostridium thermocellum*, and its expression in *E. coli* revealed that it represents a novel class of cellobiohydrolase *(14, 16)*.

The largest scale application of cellulose conversion, not surprisingly, uses the best studied cellulase, that of *Trichoderma reesei*, to produce ethanol through fermentation. A pilot plant built by Raphael Katzen Associates International, Inc. (Cincinnati, OH) is a batch-fed, simultaneous saccharification and fermentation system (SSF) *(17)* that efficiently recycles the cellulase *(7)*. Simultaneous saccharification and fermentation circumvents the glucose feedback inhibition of the *Trichoderma reesei* cellulase by having the cellulase and the yeast present in the same reaction vessel; as quickly as the glucose is produced (saccharification), it is converted to ethanol (fermentation). This plant, which is located at a pulp mill in Pennsylvania, has a feedstock input capacity of one ton per day and has produced ethanol at 80 to 90% of theoretical yield. Raphael Katzen Associates International, Inc. plans to build a larger plant with a feedstock capacity of 50 to 100 tons. Private sector activity such as this, which has taken advantage of knowledge gained from ongoing research at NREL, is a critical element for developing the technology into practical commercial use.

Microbes as Biocatalysts. Once one has the basic chemical starting material, glucose, the next step is to convert it to the desired organic chemical. The example of ethanol production was just discussed, but the potential exists for making biocatalytically a variety of other chemicals as well. The term biocatalysis can be used to describe a chemical reaction that is catalyzed either by a whole living organism (such as a microbe) or a single enzyme derived from an organism.

Metabolic Pathway Engineering. The essential feature of metabolic engineering is subversion of part of a cell's normal metabolism so that it can contribute to the production of the chemical of interest *(18, 19)*. The ideal is to obtain as high a yield as possible without killing the cell. Recombinant DNA techniques have permitted the cloning and manipulation of a great variety of genes encoding enzymes. There are numerous examples of specialty chemicals produced using *E. coli* and other microorganisms *(20)*, but there are very few examples of commodity chemicals produced by this technology. Notable examples of specialty chemicals include antibiotics, which have been in widespread production since the 1940's, and amino acids.

Aromatic Chemical Pathways. A classic example of metabolic engineering, and one which will be examined in considerable detail, comes from the work of Dr. John Frost of Michigan State University. Using *E. coli* as the biocatalyst and glucose as a feedstock, Frost and colleagues have successfully manipulated key enzymes involved in aromatic amino acid biosynthesis, and in the process discovered routes to very useful industrial chemicals. Frost focused first on 3-deoxy-D-arabino-heptulosonic phosphate (DAHP) synthase, which controls the flow of carbon into aromatic amino acid biosynthesis by catalyzing the condensation of D-erythrose 4-phosphate (E4P) and phosphoenolpyruvate. DAHP synthase was overexpressed in *E. coli* from a plasmid. In an effort to boost the levels of DAH and DAHP produced by DAHP synthase, the gene for a second enzyme, transketolase, was cloned into the same plasmid. Transketolase lies upstream of DAHP synthase and converts D-fructose 6-phosphate into E4P. In an *E. coli* mutant incapable of metabolizing DAH and DAHP, their levels were amplified 23-fold *(21)*.

The next molecule in the pathway, 3-dehydroquinate, is produced by the rate-limiting DHQ synthase. Overexpressing this enzyme from the same plasmid encoding DAHP synthase, and transketolase in another *E. coli* mutant lacking functional shikimate dehydrogenase, resulted in the conversion of 56 mM glucose to 30 mM 3-dehdroshikimate (DHS), the next intermediate *(22)*.

Frost and Draths *(23)* were unsuccessful in their attempt to continue their systematic accumulation of intermediates in the aromatic amino acid biosynthetic pathway when the three enzymes discussed above (DAHP synthase, transketolase, and DHQ synthase) were overexpressed in an *E. coli* strain lacking chorismate synthase activity. This led to the discovery that under the conditions of their assay, catechol was being produced, along with beta-ketoadipate. The *Klebsiella pneumoniae* enzymes involved in the conversion of DHS to catechol, DHS dehydratase and protocatechuate decarboxylase, were used since the corresponding *E. coli* enzymes have not yet been cloned. This discovery may have a significant industrial impact, as these two chemicals have important uses in the chemical industry. Catechol, for example, can be used to produce the flavoring vanillin, as well as L-dopa, epinephrine, and norepinephrine. Adipic acid is used in the production of nylon-6,6.

This technology has the potential to replace benzene in the manufacture of nylon-6,6. Approximately 98% of adipic acid is made from cyclohexane, which in turn is made from benzene. Benzene is flammable, highly toxic, and a known carcinogen; in contrast, glucose is non-toxic.

Frost and Draths *(24)* have also discovered a way of producing quinic acid, another important industrial chemical. By introducing the gene for quinic acid dehydrogenase from *Klebsiella pneumoniae* into an *E. coli* strain lacking DHQ dehydratase and overexpressing the trio of enzymes discussed above, Frost created an organism that cannot metabolize quinic acid. Subsequently, quinic acid can be converted in high yield to both benzoquinone and hydroquinone in a single step. Benzoquinone is used as a building block for the synthesis of a variety of organics; hydroquinone is used primarily in photography.

Frost and his group are currently engaged in scale-up development. To this end, they have eliminated the concern surrounding plasmid maintenance by integrating the plasmids into the *E. coli* chromosome. As a result, the promoter-gene constructs no longer have to be selected by using a plasmid-encoded antibiotic resistance marker, which is expensive and risky in large-scale operations.

1,3-Propanediol. Another investigator involved in the field of commodity chemicals through biotechnology is Dr. Douglas Cameron at the University of Wisconsin, Madison. Cameron's group has genetically engineered *E. coli* to produce 1,3-propanediol (1,3-PD). This was accomplished by transforming *E. coli* with genomic DNA fragments isolated from a natural producer of 1,3-PD, *Klebsiella pneumoniae*, and screening for the production 1,3-PD. The genes responsible for this activity were found to be glycerol dehydratase, which converts glycerol to 3-hydroxypropionaldehyde (3-HPA), and 1,3-PD oxidoreductase, which converts 3-HPA to 1,3-PD *(25)*. Although the recombinant strain produces less 1,3-PD than *K. pneumoniae*, it is much more amenable to genetic manipulation, which will facilitate further metabolic engineering aimed at increasing yields. A disadvantage that could also be addressed by metabolic engineering is that the recombinant strain still requires glycerol as a substrate, which is more expensive and less abundant than sugars, starches and cellulosics. Alternatively, this price differential may be minimized with the anticipated sharp increase in biodiesel fuel production, of which glycerol is a byproduct.

1,3-PD is currently produced commercially in small quantities by chemical synthesis using the toxic feedstock acrolein. Although 1,3-PD has not been produced on a large scale, there are dozens of potential uses in polymer synthesis and as a chemical intermediate *(26)*. Cameron has also been involved in studies on strains of *Clostridium thermosaccharolyticum* that produce R(-)-1,2-propanediol, a useful chiral building block in organic synthesis *(27)*.

Ethanol. Another clever use of recombinant DNA technology takes us back again to ethanol production. One disadvantage of relying on the yeast *Saccharomyces cerevisiae* for fermentation is that it can only use glucose as a substrate. Dr. Lonnie Ingram and his co-workers at the University of Florida, Gainesville, have genetically engineered the gram-negative bacteria *E. coli (28)*, *Klebsiella oxytoca (29)* and *Erwinia sp. (30)* to use both the glucose and the hemicellulose five-carbon sugars derived from lignocellulose. They accomplished this by introducing into these strains two genes from the bacterium *Zymomonas mobilis*. The encoded enzymes, pyruvate decarboxylase and alcohol dehydrogenase, are both needed to divert pyruvate metabolism to ethanol and are the only enzymes required in the ethanol pathway of *Z. mobilis*. An interesting feature of the modified *K. oxytoca* is that it naturally contains a native transport system for cellobiose and cellotriose and can metabolize these compounds further *(31, 32)*. The need for the least stable component of cellulase, beta-glucosidase, is thus obviated. These exciting advances are being developed for commercial use by BioEnergy International, a subsidiary of Quadrex Corporation.

Succinic Acid. The AFP (Alternative Feedstocks Program, U.S. Department of Energy) is focusing on succinic acid production by fermentation, with the intention of making succinic acid a commodity chemical. Workers at Argonne National Laboratory (ANL) have succeeded in selecting for a natural, high succinic acid producer, *Anaerobiospirillum succiniproducens*. The key enzymes for this metabolic pathway are well-known. After conversion of glucose to phosphoenolpyruvate (PEP), PEP carboxykinase and PEP carboxylase convert PEP to oxaloacetate. Malate dehydrogenase then converts oxaloacetate to malate, which is converted to fumarate by fumarase. Finally, fumarate reductase converts fumarate to succinate. ANL reported in 1994 that 30-fold overexpression of the *E. coli* PEP carboxylase in *E. coli* resulted in a 5-fold boost in the amount of succinic acid produced by this organism. Although less than the yield achievable with the non-engineered *A. succiniproducens* (at least 3-fold less), ANL predicts that further metabolic engineering and selection could eliminate this gap. The focus in this area, however, will probably turn to engineering *Lactobacillus* strains such as *L. gasseri*, which have a very high tolerance to organic acids and result in very high yields of succinate, on the order of 3-fold or higher (>100 g/liter) than what is currently possible with *A. succiniproducens*.

Succinic acid can be converted into a wide variety of highly useful industrial chemicals, including succinate esters and derivatives (which could serve as "green" solvents), 1,4-butanediol, tetrahydrofuran, gamma-butyrolactone, biodegradable polyesters, maleic anhydride, and others. Again, the starting material for succinic acid production is non-toxic glucose, not petroleum.

Lactic Acid. ANL also reported in 1994 that although most *Lactobacillus* strains are refractory to transformation with plasmid DNA, electroporation techniques are beginning to show positive results. This is, of course, a prerequisite for metabolic engineering. Several *Lactobacillus* strains produce very high yields of L(+)-lactic acid from glucose. Even without metabolic engineering, two of the best strains, *L. delbreuckii* and *L. helveticus*, produce a very respectable 100 grams per liter in a little over a day. ANL predicts that metabolic engineering will raise this further.

Lactic acid has been used in the food, chemical and pharmaceutical industries for years, and has the potential to achieve commodity-level status as an intermediate for oxygenated chemicals, "green" solvents, specialty chemical intermediates, and polylactic acid (PLA). PLA, which is already used in medical devices, holds great promise for increased use as a versatile, environmentally-friendly, biodegradable plastic. The L(+) enantiomer of lactic acid appears to be important for producing high quality PLA. A number of US companies have recently built development-scale plants for producing lactic acid and polymer intermediates *(33)*.

Bioprocessing. A problem associated with chemicals-through-fermentation is feedback inhibition by the product, which, of course, leads to decreased yields. Unfortunately, the concentration at which the final product is inhibitory typically is extremely dilute relative to standard organic chemical synthesis techniques. If the chemical is an organic acid, the common purification practice has been to precipitate the acid with added salt, which leads to an enormous salt waste problem. Though not toxic *per se*, its sheer mass creates a solid waste problem that is unacceptable.

Biparticle Fluidized Bed Reactors Technology. New engineering techniques have recently been developed that permit continuous removal of product from the reaction stream. A biparticle fluidized-bed bioreactor has been used to produce lactic acid in this way *(34)*. In the bioreactor is placed *L. delbreuckii* immobilized on alginate beads. Larger polyvinyl pyridine (PVP) beads, which act as organic acid adsorbent, are fed through the top of the reactor, while a liquid stream is fed in through valves at the bottom of the reactor, creating an upward current. The large sorbent particles have a Stokes' settling velocity that permits them to descend through the current generated by the feed stream and pass through a particle removal valve at the bottom of the reactor. The smaller biocatalyst beads, unable to settle against the liquid feed stream, are retained in the bed. After the sorbent beads exit the reactor, the lactic acid is removed from them. The stripped beads are fed through the top of the reactor again, and the cycle is repeated. Thus, continuous fermentation can take place, and lactic acid is removed on the PVP beads, relieving the feedback inhibition problem and effectively concentrating the fermentation product.

A similar system could be applied to research being conducted in Dr. George Tsao's laboratory at Purdue University in which the fungus *Rhizopus oryzae* is used to produce lactic acid *(35)*. Like *Lactobacillus*, *Rhizopus oryzae* produces the L(+) enantiomer from glucose, but offers the additional advantage of producing L(+)-lactic acid from xylose. Given that xylose is the primary constituent of hemicellulose, this work may be relevant to researchers who are investigating the application of simultaneous saccharification and fermentation to the production of (L+)-lactic acid from lignocellulosics *(36)*.

Two-Stage Electrodialysis. Progress in using electrodialysis techniques for recovering organic acids has been reported recently. A two-stage electrodialysis primary purification technique has been developed by ANL. In the first part of this process, the fermentation broth is subjected to desalting electrodialysis, during which the crude organic salts are concentrated to approximately 3N and separated from impurities, including protein, of which about 90% are removed at this stage. The partially purified and concentrated sodium salt of the organic acid (e.g., sodium lactate) then undergoes water-splitting electrodialysis *(37)* where the free organic acid (e.g., lactic acid) is separated from the sodium ion. At the same time, water is separated into hydrogen and hydroxyl ions. The sodium and hydroxyl ions then form sodium hydroxide, which is captured as a useful byproduct, and the lactic acid is subjected to a secondary purification step. Using this two stage primary purification system, the generation of salt waste is eliminated.

Isolated Enzymes and Other Biocatalysts. As mentioned above, isolated enzymes fall under the biocatalyst domain. Purified enzymes probably will not play as large a role as microorganisms in commodity chemicals production from biomass, largely because of the impracticality of preparing all of the many enzymes necessary to convert glucose to a useful chemical. Purified enzymes will have an impact, however, in the pharmaceutical and fine chemicals industries, where a critical, value-added downstream step needs to be performed.

Enzymes in Supercritical Fluids. The effects of organic solvents and supercritical fluids on enzyme activity, stability, and specificity have been studied. Having evolved in aqueous systems, enzymes are clearly in an unnatural environment when in these solvents. Interestingly, some enzymes are active in nonaqueous solvents, and the activity can be altered dramatically in supercritical fluids *(38)*. One of the reactions studied involves the use of *Candida cylidracea* lipase in catalyzing the alcoholysis reaction between methyl methacrylate and ethylhexanol. This reaction and similar ones can be used to produce Plexiglas (methylmethacrylate), controlled drug release matrices (hydroxymethacrylates), and contact lenses (hydroxyacrylates). The activity of the lipase can be fine-tuned simply by altering the pressure of the supercritical fluid *(39, 40)*.

Extremozymes. Of relevance to the work surrounding enzymes in supercritical fluids is the growing interest in enzymes from microbes known as extremophiles *(41, 42)*. As the name implies, extremophiles are bacteria that have evolved under extreme conditions of high heat or high salt concentrations. One of the best known enzymes isolated from an extremophile is the Taq DNA polymerase from *Thermus aquaticus*, which has made the polymerase chain reaction a much more efficient and less labor-intensive technique. Taq polymerase is being joined by a host of other polymerases isolated from other bacteria that are even more resistant to high temperature. These thermophiles have the full complement of "housekeeping" enzymes and may have novel enzymes, all of which presumably have been optimized by nature to function at 90°C or warmer. Enzymes from halophilic bacteria require a high salt concentration for activity, which may have relevance for industrial reactions requiring the use of organic solvents, since high salt and organic solvents share the property of dehydrating enzymes. Although extremophile research is at a very early stage, it will likely have an impact on industrial processes involving enzymes.

Catalytic Antibodies. Another area of intense activity is the field of catalytic antibodies *(43, 44)*. In a short period of time, antibodies have been engineered to catalyze more than fifty diverse types of chemical reactions. Animals are injected with a hapten that represents the transition state analog for the reaction of interest, and the extremely high diversity of the immune system yields antibodies specific for the analog.

Catalytic antibodies are capable of catalyzing reactions normally carried out by enzymes, albeit usually with much lower efficiency. Interestingly, an x-ray crystallographic analysis of the structure of a catalytic antibody that mimics chorismate mutase showed that it uses essentially the same mechanism to carry out the reaction *(45)*. A similar finding was made for a catalytic antibody with a serine protease active site *(46)*. Both of these observations are fascinating because while enzymes evolved over millions of years, the catalytic antibodies were generated in only a matter of weeks.

Researchers have recently focused on developing catalytic antibodies for reactions that are difficult to achieve with existing chemical methods *(47)*, an area where real value can be added. One of the goals in the field of catalytic antibodies is to develop antibodies that carry out novel chemical reactions not known to have enzymatic counterparts in nature. Although the potential for this does exist, it is a long way from being realized.

Ribozymes. Perhaps further removed than catalytic antibodies are from typical industrial catalysts are biocatalysts that are not protein-based. Naturally occurring ribonucleic acids called ribozymes are capable of carrying out RNA cleavage reactions with a high degree of specificity *(48)*. Artificial ribozymes have also been engineered to cleave desired RNA targets *(49, 50)*. Recently, researchers created a remarkable, novel ribozyme that acts as a polynucleotide kinase (PNK), an activity that is possessed in nature only by protein enzymes. From pools of random sequence RNA molecules, Sassanfar and Szostak *(51)* selected RNA molecules that were capable of binding ATP. Lorsch and Szostak *(52)* then used this motif as a core, surrounded by random RNA sequences, and in a sophisticated but simple screen, successfully selected for RNA molecules that showed PNK activity. Thus, for the first time, ribozymes have been created that carry out a reaction other than RNA cleavage. There is no reason that this could not be extended to other reactions, as well.

Discussion

US Biomass Abundance. An important point is that, in contrast to petrochemicals, the US has more than enough of its own raw materials in the form of biomass to satisfy its organic chemical needs in terms of total mass. In fact, using 1988 numbers, Leeper et al. *(2)* conservatively estimate that 540 billion pounds of organic chemicals could be produced by bioconversion, a figure that represents more than 170% of 1991 US organic chemicals production. The two primary advantages from an environmental viewpoint are that, in contrast to petroleum or natural gas, biomass is nontoxic, and is completely renewable. Eliminating U.S. dependence on foreign sources of oil for organic chemicals is also an obvious advantage from a national security and long-range economic point of view.

Economics. An important factor in the chemical industry's adoption of biomass conversion is cost. Advances in all areas of biotechnology will impact on the cost of the entire process and will decrease the barrier to entry. The apparent lower current cost of petrochemicals production should not suffice as a reason not to develop a young but viable technology. Again, arguments that petrochemicals production costs less are specious, given the high and rising costs of waste treatment, waste disposal, and regulatory compliance. Assistance from the federal government will be required in helping to establish the necessary infrastructure for an agri-based chemical industry, as it did for the petrochemical industry, which today enjoys the benefits of amortization.

The Importance of Recombinant DNA Technology and Scale-Up. For both lignocellulose utilization and metabolic engineering, recent progress has depended on recombinant DNA technology and future progress will continue to rely heavily on this technology. Various opportunities exist for using recombinant DNA technology to maximize product yields. For example, cloning the *Vitreoscilla* hemoglobin gene into *E. coli* improves growth characteristics and product yield, and may prove an effective strategy for oxygen-starved aerobic fermentations *(53, 54)*. Recombinant DNA technology can continue to be used to introduce enzymes from foreign species and to increase yields by altering the promoter regions of genes to enhance the levels of key enzymes, or by making (or simply selecting for) mutations within coding regions that may increase specific activities or stability with respect to temperature, solvent, pH, pressure and electrolytes. Experience has shown that despite the power of genetic engineering, progress can be slow, especially given the complexities of metabolic pathways and how little is understood about enzyme structure and function. Frequently updated and sophisticated computer databases will be necessary to leverage the vast amount of information that will flow from the Microbial Genome Initiative and the increased activity in metabolic engineering research.

For lignocellulose utilization, fermentation and processing, scale-up is an obvious necessity. Scaling up usually turns out to be more complicated and refractory than anticipated. Time, money, ingenuity and well-trained, interdisciplinary workers are necessary to make the transition to a level of production that would be appealing to industry.

Conclusion

The use of biotechnology for the production of specialty chemicals, such as those used by the pharmaceutical industry to make chiral drugs, is a reality currently being pursued by many companies. A significant advantage that biocatalysts offers is high efficiency and high selectivity. Although the cost of using biocatalysts today is also high, this cost can be absorbed by the high selling price that the final product (e.g., a drug) commands on the market. The use of biocatalysts can have an even greater environmental impact in the production of commodity organic chemicals because of the sheer volume of toxic chemicals generated from the current use of fossil feedstocks. Unfortunately, unlike specialty chemicals, profit margins in commodity chemicals production are very narrow; processes must therefore be as economical as possible to compete with the lower up-front cost of well-entrenched petrochemicals. Currently, this is not the case.

The Future of Biotechnology

Given that current economics favor petrochemicals over biomass chemicals, and that petroleum will be available for decades, chemicals generated from biomass will not take the industry by storm. Instead, it will slowly and steadily make inroads in niche markets where economics are in its favor. There will come a day in the next century when petroleum reserves are depleted. Some predict that it will be more cost-effective to replace petroleum with coal than with biomass *(55)*, and that the fermentation industry will simply continue to produce traditional fermentation products such as amino acids, antibiotics, and citric acid *(56)*. However, this ignores waste treatment, waste disposal, and regulatory compliance costs involved in using any fossil feedstock to produce organic chemicals. In the future these cost may very likely be prohibitive, in which case biomass will emerge as the only logical alternative feedstock. What is required today is increased research and development in all of the areas described in this chapter, so that the technology will be in place to fulfill the needs of the future. The federal government, the chemical industry and the academic community each have a critical role to play in fulfilling these needs.

Literature Cited

(1) Morris, D.; Ahmed, I. *The Carbohydrate Economy: Making Chemicals and Industrial Materials from Plant Matter*; Institute for Local Self-Reliance: Washington, DC, **1992**.

(2) Leeper, S.A.; Ward, T.E.; Andrews, G.F. *Production of Organic Chemicals Via Bioconversion: A Review of the Potential;* Report No. EGG-2645; U.S. Department of Energy: Washington, D.C., **1991**.

(3) Stryer, L. *Biochemistry*; W.H. Freeman and Company: San Francisco, 1981.

(4) Wyman, C.E.; Goodman, B.J. *Near Term Application of Biotechnology to Fuel Ethanol Production from Lignocellulosic Biomass*; NIST GCR 93-63; U.S. Department of Commerce: Washington, D.C., **1993**.

(5) *Alternative Feedstocks Program Technical and Economic Assessment: Thermal/Chemical and Bioprocessing Components*; Bozell, J.J.; Landucci, R., Eds.; Office of Industrial Technologies, US Department of Energy: Washington, D.C., **1993**.

(6) Eveleigh, D.E.; Bok, J.; El-Dorry, H.; El-Gogary, S.; Elliston, K.; Goyal, A.; Waldron, C.; Wright, R.; Wu, Y.-M. *Appl. Biochem. Biotechnol.*, in press.

(7) Katzen, R.; Monceaux, D.A. *Appl. Biochem. Biotechnol.*, in press.

(8) Sandhu, D.K.; Bawa, S. *Appl. Biochem. Biotechnol.* **1992**, *34/35*, pp. 175-183.

(9) Divne, C.; Stahlberg, J.; Reinikainen, T.; Ruohonen, L.; Pettersson, G.; Knowles, J.K.C.; Teeri, T.T.; Jones, T.A. *Nature* **1994**, *265*, pp. 524-528.

(10) Waldron, C.R.; Becker-Vallone, C.A.; Eveleigh, D.E. *Appl. Microbiol. Biotechnol.* **1986**, *24*, pp. 477-486.
(11) Wright, R.M.; Yablonsky, M.D.; Shalita, Z.P.; Goyal, A.K.; Eveleigh, D.E. *Appl. Environ. Microbiol.* **1992**, *58*, pp. 3455-3465.
(12) Johnson, E.A.; Sakajoh, M.; Halliwell, G.; Madia, A.; Demain, A.L. *Appl. Environ. Microbiol.* **1982**, *43*, pp. 1125-1132.
(13) Lamed, R.; Setter, E.; Bayer, E.A. *J. Bacteriol.* **1982**, *156*, pp. 828-836.
(14) Wang, W.K.; Kruus, K.; Wu, J.H.D. *Appl. Microbiol. Biotechnol.*, in press.
(15) Wu, J.H.D. In *Biocatalyst Design for Stability and Specificity*; Himmel, M.E.; Georgiou, G., Eds.; ACS Symposium Series; American Chemical Society: Washington, D.C., **1993**, *516*, pp. 251-264.
(16) Wang, W.K.; Kruus, K.; Wu, J.H.D. *J. Bacteriol.* **1993**, *175*, pp. 1293-1302.
(17) Blotkamp, P.J.; Takagi, M.; Pemberton, M.S.; Emert, G.H. *Enzymatic Hydrolysis of Cellulose and Simultaneous Fermentation to Alcohol.*; AIChE Symp. Ser.; **1978**, *74*, pp. 85-90.
(18) Bailey, J.E. *Science* **1991**, *252*, pp. 1668-1675.
(19) Stephanopoulos, G.; Vallino, J.J. *Science* **1991**, *252*, pp. 1675-1681.
(20) Cameron, D.C.; Tong, I.-T. *Appl. Biochem. Biotechnol.* **1993**, *38*, pp. 105-140.
(21) Draths, K.M.; Frost, J.W. *J. Am. Chem. Soc.* **1990**, *112*, pp. 1657-1659.
(22) Draths, K.M.; Frost, J.W. *J. Am. Chem. Soc.* **1990**, *112*, pp. 9630-9632.
(23) Draths, K.M.; Frost, J.W. *J. Am. Chem. Soc.* **1991**, *113*, pp. 9361-9363.
(24) Draths, K.M.; Ward, T.L.; Frost, J.W. *J. Am. Chem. Soc.* **1992**, *114*, pp. 9725-9726.
(25) Tong, I-T.; Liao, H.H.; Cameron, D.C. *App. Environ. Microbiol.* **1991**, *57*, pp. 3541-3546.
(26) D.C. Cameron, personal communication.
(27) Cameron, D.C.; Cooney, C.L. *Biotechnology* **1986**, *4*, pp. 651-654.
(28) Ingram, L.O.; Conway, T.; Clark, D.P.; Sewell, G.W.; Preston, J.F. *Microbiol.* **1987**, *53*, pp. 2420-2425.
(29) Ohta, K.; Beall, D.S.; Mejia, J.P.; Shanmugam, K.T.; Ingram, L.O. *App. Environ. Microbiol.* **1991**, *57*, pp. 2810-2815.
(30) Beall, D.S.; Ingram, L.O. *J. Indust. Microbiol.* **1993**, *11*, pp. 151-155.
(31) Al-Zaag, A. *J. Biotechnol.* **1989**, *12*, 79-86.
(32) Wood, B.E.; Ingram, L.O. *App. Environ. Microbiol.* **1992**, *58*, pp. 2103-2110.
(33) Shih-Perng Tsai, personal communication.
(34) Davison, B.H.; Thompson, J.E. *Appl. Biochem. Biotechnol.* **1992**, *34/35*, pp. 431-439.
(35) Yang, C.W.; Lu, Z.; Tsao, G.T. *Appl. Biochem. Biotechnol.*, in press.
(36) Padukone, N.; Schmidt, S.L.; Goodman, B.J.; Wyman, C.E. *Appl. Microbiol. Biotechnol.*, submitted.
(37) Bolton, H.R. *J. Chem. Tech. Biotechnol.* **1992**, *54*, pp. 341-347.
(38) Russell, A.J.; Beckman, E.J.; Chaudhary, C.K. *Chemtech*, **1994** (March), pp. 33-37.

(39) Kamat, S.; Barrera, J.; Beckman, E.J.; Russell, A.J. *Biotechnol. Bioengin.* **1992**, *40*, pp. 1589-166.
(40) Kamat, S.V.; Iwaskewycz, B.; Beckman, E.J.; Russell, A.J. *Proc. Natl. Acad. Sci.*, USA 90, **1993**, pp. 2940-2944.
(41) *Science* **1994**, *265*, pp. 471-472
(42) *The Scientist* **1994** (May 30), pp. 14-15
(43) Lerner, R.A.; Benkovic, S.J.; Schultz, P.G. *Science* **1991**, *252*, pp. 659-667.
(44) Stewart, J.D.; Liotta, L.J.; Benkovic, S.J. *Acc. Chem. Res.* **1993**, *26*, pp. 396-404.
(45) Haynes, M.R.; Stura, E.A.; Hilvert, D.; Wilson, I.A. *Science* **1994**, *263*, pp. 646-652.
(46) Zhou, G.W.; Guo, J.; Huang, W.; Fletterick, R.J.; Scanlan, T.S. *Science* **1994**, *265*, pp. 1059-1064.
(47) Schultz, P.G.; Lerner, R.A. *Acc. Chem. Res.* **1993**, *26*, pp. 391-395.
(48) Cech, T.R. *Science* **1987**, *236*, pp. 1532-1539.
(49) Dropulic, B.; Lin, N.H.; Martin, M.A.; Jeang, K.-T. *J. Virol.* **1992**, *66*, pp. 1432-1441.
(50) Siuod, M.; Natvig, J.B.; Forre, O. *J. Mol. Biol.* **1992**, *223*, pp. 831-835.
(51) Sassanfar, M.; Szostak, J.W. *Nature* **1993**, *364*, pp. 550-553.
(52) Lorsch, J.R.; Szostak, J.W. *Nature* **1994**, *371*, pp. 31-36.
(53) Khosla, C.; Bailey, J.E. *Mol. Gen. Genet.* **1988**, *214*, pp. 158-161.
(54) Khosravi, M.; Webster, D.A.; Stark, B.C. *Plasmid* **1990**, *24*, pp. 190-194.
(55) Hinman, R.L. *Biotechnology* **1991**, *9*, pp. 533-534.
(56) Hinman, R.L. *Chemtech* **1994** (June), pp. 45-48.

RECEIVED December 19, 1995

Educational Tools

Chapter 16

Teaching Alternative Syntheses: The SYNGEN Program

James B. Hendrickson

Department of Chemistry, Brandeis University, Waltham, MA 02254–9110

The generation of alternative, environmentally benign, synthesis routes depends on having a rigorous and logical protocol for synthesis generation. This would also constitute a reasonable basis for teaching synthesis design systematically. Such a protocol is developed here as an outline for a teaching course in synthesis design and is implemented in the SYNGEN program. The value of such a program lies in its ability to offer a number of alternative synthesis routes which could be environmentally less hazardous or polluting, on the grounds that it is better to minimize pollution in the design first than in the use later.

Thousands of organic chemicals are synthesized annually on an industrial scale, and their manufacture can often lead to environmental problems. If alternative syntheses that create fewer hazardous wastes and less pollution could be found, a number of these problems could be solved. No one would claim that the synthesis routes currently in use are the only ones possible, or even that they know them to be the best routes. However, organic chemistry has traditionally not provided a logical protocol for the systematic design of synthesis routes to any target molecule.

If there were such a protocol, other routes to any industrial synthesis target could be systematically explored, and their relative impacts on the environment examined. Also, if there were such a protocol, it could be incorporated into the teaching of organic synthesis to show the variety of different synthesis routes possible to any target molecule. At the same time it would point out their differences in terms of environmental impact.

Indeed it is easy to show that the number of possible synthesis routes to target molecules of even modest complexity is usually enormous. One reason that systematic synthesis design protocols have not been devised until recently is simply that the great capacity of the computer has not been available to master the vast

0097–6156/96/0626–0214$12.00/0

number of structures and reactions involved in generating and assessing all these routes.

This discussion will develop stepwise a logic of synthesis design that can become the basis for a course in teaching synthesis design to chemists and chemistry students. The logic also serves to direct a computer program, SYNGEN, which seeks to generate all the best synthesis routes systematically. The course and the program emphasize the variety of synthesis routes possible and so encourage chemists to think about alternative routes and to choose the environmentally benign ones.

Computer Generation of Synthesis Routes

The first publications to apply a computer system to the problem of synthesis design came from Corey (1) and Bersohn (2) a quarter of a century ago. Provided with an internal database of organic reactions, these programs then generated all the last-step reactions to the target structure from intermediate molecules one step back. Each of these intermediates then became a subtarget and the process was repeated on each one to generate the intermediates two steps back from the original target. This process can be systematically repeated until the routes being generated result in acceptable starting materials. Since this beginning many programs have been written (3), all of which exploit the ability of computers to generate systematically all reactions back from the target.

The problem with this procedure is the number generated. For a target structure of reasonable complexity, one may easily derive some 30 possible reactions for the last step leading to the target. Since most of the intermediate compounds so generated will still have comparable complexity, 30 ways may be expected to make each of these. In two steps 900 routes are already generated, in five steps 30^5, or 24 million, routes; and in many cases acceptable starting materials will not yet have been generated. Not uncommonly, even industrial syntheses have ten or more steps, and so the possible variety has grown to astronomical dimensions. The problem then is not so much the generation of all possible routes as it is the selection of which few are the best.

In most of the present programs (3) it is left to the operator to decide which of the synthesis routes so generated will be kept. Thus the operator does the main intellectual work of finding the best routes. However, the only way he can do this is to prune out most intermediates when they first appear, even at the first step, and so he will never know which of those might have led to a better or shorter synthesis route. To arrive at a more helpful synthesis protocol the computer must do more of the intellectual work, and must apply more logic in order to prune these huge numbers very stringently, focusing the search more sharply for just a few best synthesis routes.

The generation of all reactions in each synthesis route, as described above, results in an enormous "synthesis tree" of intermediate structures and reactions, rooted at the right end on the target structure and growing to the left toward starting material structures. The synthesis tree may be represented as a kind of "process graph", in Figure 1, showing the lines as reaction steps and the points as

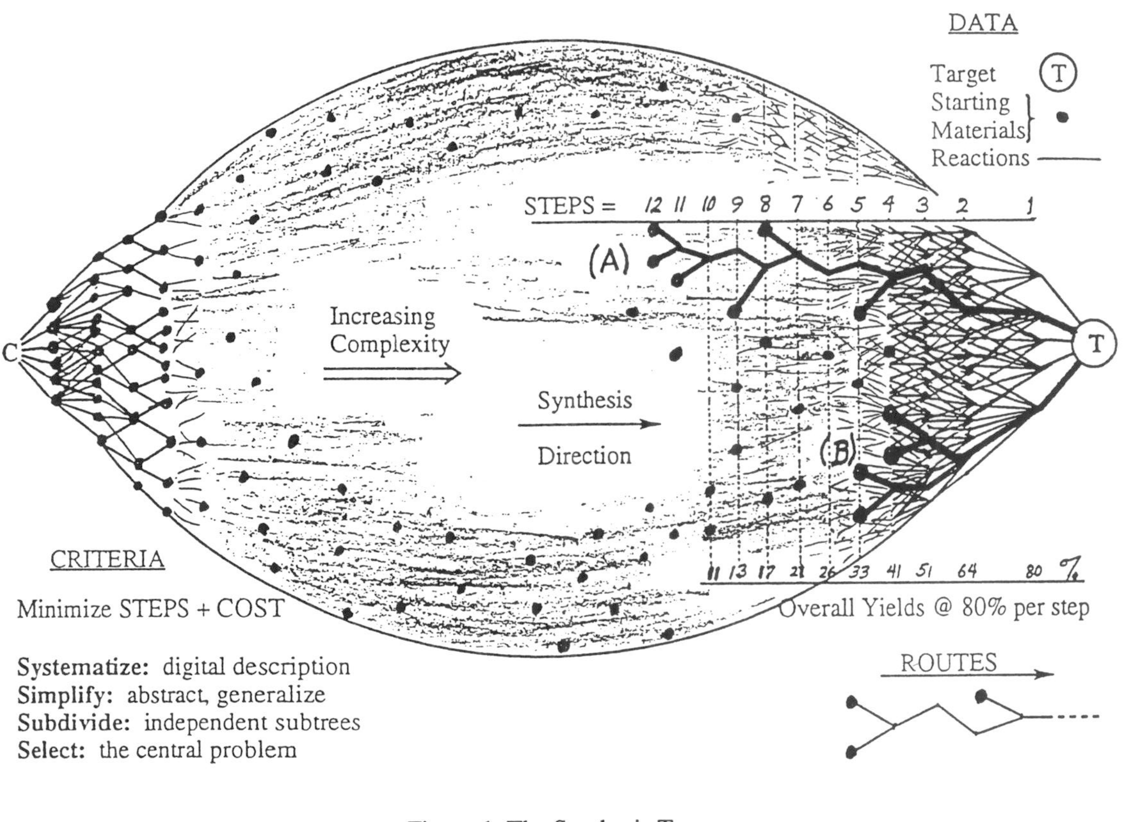

Figure 1. The Synthesis Tree

chemical compounds: the target (T); the starting materials shown as heavy dots; and the many intermediates as the points which end each reaction line. All the reactions are directed to the right - toward the target - and two lines meeting to the right signify the joining of two compounds to form an intermediate. The compounds in the tree become less and less complex toward the left side, ultimately culminating in the simplest one-carbon starting materials at extreme left, designated simply as "C". Any routes which start at C must build the target in one-carbon increments and will therefore be seen as the longest and least efficient routes of all.

An individual synthesis route is itself a tree, consisting of the several starting materials, linked through their joining reactions and intermediates to the target structure. Each route is buried among millions of others in the whole synthesis tree. The graph of the steps comprising any one route is called a *synthesis plan.* In Figure 1 two sample routes, or plans, are shown in boldface: A with 12 steps from 6 starting materials and B with only 6 steps and 4 starting materials.

For the manipulation of such a huge tree, several main procedures which have proved useful in other fields can also be useful here.

1. Simplify the variety and number of items (structures and reactions) that must be manipulated in the search by generalizing and abstracting them, coalescing lesser distinctions into broader families. When only a few have survived the applied criteria, those few may then be refined back to full chemical detail.

2. Systematize the definition of structures and reactions into a digital format for facile computer manipulation. This provides for sharp definition of the generalized families, and this in turn focuses the simplification of the field. Since all possibilities can then be defined as mathematical combinations, this also assures that no possible routes will be missed.

3. Subdivide the whole synthesis tree into independent subtrees, each of which can more easily be encompassed, generated, and evaluated.

4. Select finally the few best synthesis routes. This may be done most simply by putting the remaining routes in an order based on some quantitative measure, such as cost and/or environmental impact, to direct preferences.

Criteria. A critical focus for any search of this huge synthesis tree must be a definition of what are the "best" synthesis routes. We might start by imagining that the chemist who needs to make something will prefer a synthesis of one step in 100% yield from a cheap starting material source! This preference implies minimizing not only *effort*, expressed by the number of reaction steps, and *cost* of starting materials, but also byproducts, implied in *yields*. However, for reactions which have not been carried out before (regardless of their precedent), the experience of predicting yields has always been inaccurate to the point of uselessness.

The only really reliable criterion to apply then is the number of steps. The best syntheses are the shortest. Calculation of the overall cost of any route will show that the starting material costs amplified by yield loss are therefore a function of the number of steps; but since the yield figure employed in such a calculation is unknown, the value used must by default be just an average number. This will be the 75-80% average of all reported reactions in reaction databases. These overall yields, at 80% per step, are shown in Figure 1, declining significantly the farther the starting materials are from the target and consequently much enlarging the initial starting material costs.

Starting Materials. At the other end of the synthesis tree from the target structure are the starting materials (Figure 1). Carried to its ultimate limit, the backwards generation of intermediates will arrive at the simplest, one-carbon starting materials (C in Figure 1), creating synthesis routes "from coal, air and water", as such paths were described in the 19th century. This of course has the consequence that every carbon-carbon bond in the target must be made, and so such routes will generally have at least as many steps as there are skeletal bonds in the target.

Now that the chemical industry manufactures, i.e., synthesizes, larger starting material molecules from petroleum and plant matter, the synthesis routes from these to target molecules are much shorter. This in turn implies that a catalog of these thousands of available starting materials must be provided to the generation process, so that the computer will know when to stop, as it works backwards from the target.

Furthermore, if it were possible to plot the entire synthesis tree for a desired target molecule and mark the catalog starting materials on it, the result would certainly be that the shortest routes are those in which the starting materials are closest to the target in the synthesis tree. Therefore, it will be sensible to create a generation process that can locate these closest precursors and direct our search for best routes back just to these directly, for such routes must be the shortest ones.

Starting materials are almost always smaller than the target molecule and several, typically 3-6 carbons, must be linked together to form it, as seen in Figure 1. In fact starting materials are usually surprisingly small: in a survey of many syntheses it was found that the average starting material size for non-aromatic targets was *only three carbons* actually incorporated into the target. On the other hand the 6000 starting materials in the SYNGEN catalog show much functional variety up through about *five* carbons in size, suggesting that it should be possible to find routes more efficient than the average, with bigger starting materials and therefore fewer joining steps.

Focus on Skeleton

Since the work of the synthesis steps is to make the target bonds, it is important to see molecules not as assemblies of atoms but rather as networks of bonds, especially as links of σ-bonds among non-hydrogen atoms. If the ensemble of starting material bonds is mapped onto the target structure, the missing target bonds are those that must be made. To minimize steps in the synthesis route then, it is necessary to minimize these missing bonds.

Most non-hydrogen atoms are carbons, and they have a greater valence than the other atoms commonly present (e.g., O, N, Hal). Therefore, most of the bonds in the molecular network are carbon-carbon σ-bonds, a kind of linked backbone framework within the structure. It is crucial in the development of synthesis design to make a sharp distinction in any structure between its *skeleton,* composed of the σ-bonds between carbons, and at particular sites on that skeleton its attached *functional groups,* composed both of π-bonds between carbons and of attached bonds to heteroatoms (σ- or π-), i.e., the reactive centers. Chemists generally tend to think of a synthesis in terms of these functional groups, and their transformation in successive reactions. In contrast it is important to stress the underlying patterns of *skeleton elaboration* when comparing synthesis routes.

In logical terms, *synthesis is a skeletal concept.* This derives from the observation that starting materials are virtually always smaller than the target molecule. Therefore synthesis consists of joining together several small starting molecules to make a larger target. These must be joined by constructing skeleton bonds. If the skeleton bonds of the several starting materials are mapped onto the target skeleton, the missing bonds are those that must be constructed.

For any synthesis route there may be defined a *bondset* on the target skeleton, i.e., as that set of bonds which are the ones to be constructed from the given starting material skeletons (4). Alternatively, if any bondset is marked onto a target skeleton, it defines the starting skeletons directly, by deleting those bondset bonds. This is illustrated with a set of actual starting materials for a commercial synthesis of estrone in Figure 2 (sequence A). Stripped to their skeletons, these compounds are shown below it as sequence B, in which the bondset is marked on the target skeleton with boldface bonds. Furthermore, the bondset bonds in Figure 2 are numbered with the order in which they are constructed in the synthesis route.

This constitutes an ordered bondset, which is a critical summary of the whole synthesis route, for it defines not only the starting skeletons but also the sequence in which the starting materials are linked together. Sequence B is read as "fragment A is joined to fragment B by bond 1, then cyclized for bond 2; fragment C is then joined to fragment D at bond 3 and the two resultant pieces are now joined first by bond 4, then cyclized at bond 5." This skeletal analysis can be made for any synthesis. Indeed it is a very simple but very cogent way to assess the skeletal essentials of any given synthesis..

Assembly of the Skeleton. We can conveniently diagram this sequence in the ordered bondset as a *skeletal assembly plan,* shown below the bondset in Figure 2 as graph C. This is a graph of the synthesis process which is derived directly from the ordered bondset. It shows the lettered starting skeletons linked by lines for the reactions which join them, constructing the bondset bonds. The numbers represent the bonds constructed at each step as well as the order of the reactions which construct them. Such an assembly plan is also a reduced form of a whole family of full synthesis plans, all of which make the same sequence of skeletal bonds from differently functionalized starting materials and intermediates. The synthesis plan B in Figure 1 is in fact one of the whole routes incorporated in the skeletal assembly plan of Figure 2 (graph C).

This analysis has an important consequence for synthesis design. Noted above was the need for simplification of the search space in examining the whole synthesis tree. Such a major simplification is to delete the functional groups altogether and start by dissecting just the skeleton of the target into possible

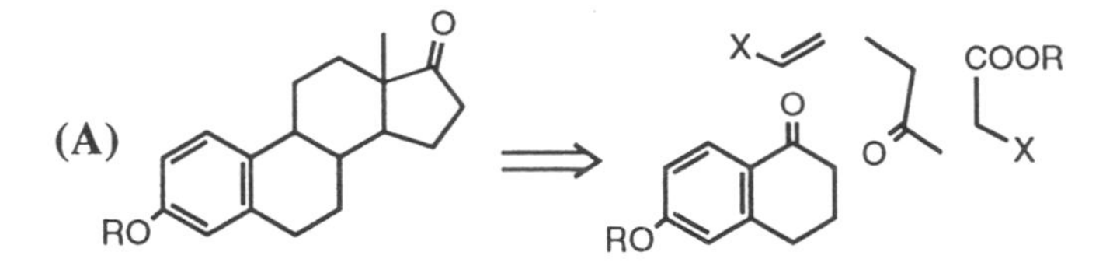

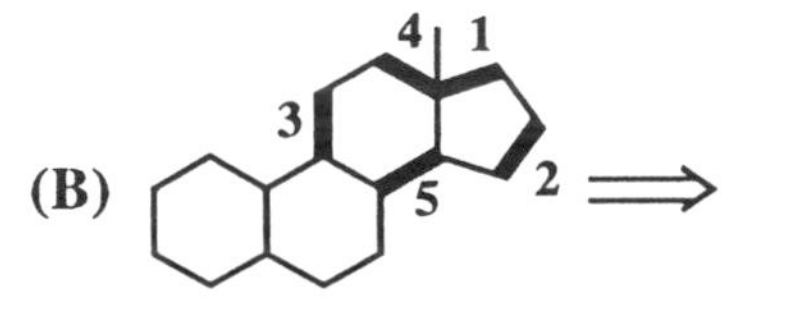

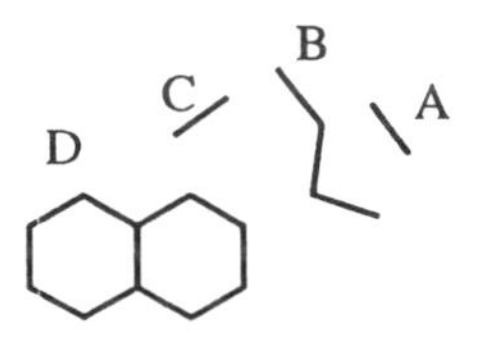

BONDSET

(C) Skeletal Assembly Plan:
(from ordered bondset)

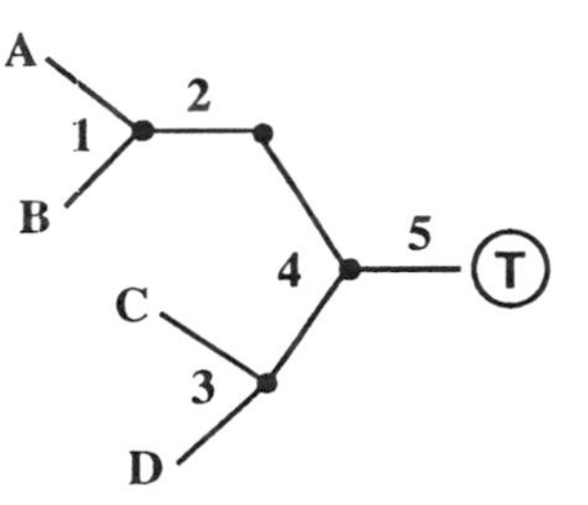

The simplest overview of any synthesis is its ordered bondset.

Figure 2. Target Dissection and Assembly

starting skeletons. Although there are tens of thousands of available starting materials, there are only 13 acyclic starting skeletons of six carbons or less and not many more cyclic ones.

This simplification focuses simply on finding the best ways to assemble the target skeleton, which means deriving bondsets on it, and so creating the skeletal assembly plans which are derived from those bondsets. Since defining a bondset defines the starting skeletons (sequence B in Figure 2), the only acceptable bondsets will be those in which the starting skeletons are all found in the catalog of available starting materials.

Any bondset represents a family of fully articulated synthesis plans, all of which construct just those skeletal bonds of the bondset from various sets of starting materials, which all have the same set of skeletons but different functional groups. Each bondset is separate from any other and its reactions do not overlap with any other. Therefore, these bondsets are all independent subtrees within the overall synthesis tree. This gives structure to the synthesis tree and affords the needed basis for subdivision of the tree search into more manageable units for examination. Each bondset may separately be fleshed out with the functional groups necessary to initiate the sequence of bond constructions dictated by its corresponding assembly plan.

The task now is to determine which skeletal bonds in the target to construct and in what order. There are indeed many ways to dissect a target skeleton into bondsets. This can be seen as a combinatoric problem, for there are $b!/n!(b-n)!$ ways to cut n bonds from a target with b bonds. For the 21 bonds of estrone cut five times as in Figure 2, there are 20349 bondsets, but there are 5! or 120 different orders for each one; therefore there are 2.44 million ordered bondsets. On the other hand, if the starting skeletons were to average just three carbons and be acyclic, there would have to be $n = 9$ bonds cut, and this will result in 10^{11} assembly plans, i.e., ordered bondsets.

Fortunately, some simple guidance of economy can vastly reduce the number of acceptable choices to those with the fewest steps, i.e., the bondsets with the fewest skeletal bonds to make. The consequence of this is that the starting materials must already be as large as possible, i.e., with many of its bonds already constructed. This suggests limiting the number of starting materials accepted, and the protocol adopted below seeks no more than four. This is usually reasonable since starting materials for a target of 20 carbons would then average five carbons, and the structural variety of available starting materials in any catalog sharply declines only above five carbons. Within this limit one should find ample bondsets and routes, and not have to look at longer routes. More starting materials than four will require more bonds to make, hence more steps in the routes.

The dissecting protocol is designed to create syntheses which are convergent, and so more efficient (5). The target skeleton is first cut all ways into two pieces. This will cut one acyclic bond or two bonds in any one ring. This cutting in two must be done three times to arrive at four starting skeletons. This procedure automatically creates a convergent synthesis route and implies a maximum of six bonds cut, if all are in rings; the estrone bondset in Figure 2 cuts only five.

Furthermore, an acceptable bondset must also find all four derived skeletons in the catalog of available starting materials.

The generation in this way of all bondsets with four acceptable starting skeletons is easily systematized and also constitutes a drastic reduction in the total number possible. With a minimum starting material size of two carbons and no cuts in the aromatic ring, for example, there are only 11 ways to make the first cut of the estrone skeleton into two pieces (B in Figure 2).

This focus on skeleton sharpens and simplifies the search in the synthesis tree, and makes it possible to define rigorously all the possible bondsets that fit the criterion of economy - the fewest skeletal bonds to make. A critical summary of any synthesis is just the sequence of those target skeleton bonds which are made, in linking the starting skeletons to assemble the target skeleton. This is its ordered bondset and defines the skeletal assembly plan which underlies the whole synthesis.

Construction Reactions. The important dichotomy in any structure between the skeleton and the functional groups has a parallel dichotomy in reactions. Reactions may be divided between *constructions,* reactions which create the skeletal bonds, and *refunctionalizations*, which alter the functional groups without affecting the skeleton. Constructions, therefore, are defined as reactions which make carbon-carbon σ-bonds.

The full sequence of steps in a synthesis will be not only the central constructions of the bondset, but also the refunctionalization reactions which are used to prepare the functional groups: (a) for their role in construction; and (b) for final conversion to the target structure after the last construction. In an average published synthesis there are actually twice as many refunctionalization steps as construction steps.

The central, obligatory reactions then are the construction reactions since the several starting materials must be joined. In principle, however, the refunctionalizations are not needed at all as long as the sequence of constructions produces the correct functional groups in the target at the end of the synthesis. This perception leads to a stringent criterion of economy: to seek only *ideal syntheses,* those consisting only of sequential construction steps with no refunctionalization steps at all, since these must be the shortest routes.

In an ideal synthesis the chosen starting materials have the correct functional groups to initiate the first construction joining two of them; this will then leave the correct functional groups for the next construction, and so forth, until the target skeletal bonds have all been made, and the last construction has left the correct functional groups on the target as well. In other words the ordered bondset, i.e., the assembly plan, is carried out as just a succession of construction reactions from given starting materials to the correctly functionalized target molecule.

Such ideal syntheses are rare but they serve as a very stringent focus on the best routes from the synthesis tree. If any ideal routes are found, they must be the shortest routes possible.

Descriptions of Structures and Reactions

Considering only skeletons and the ways to assemble them greatly simplifies and focuses the synthesis tree. The next step is to fill in the functional groups and the changes they undergo in the reactions. This will require a description system which compactly generalizes and abstracts the functional groups, coalescing trivial detail without losing too much chemical significance. The system must be rigorous in definition to include all possibilities, and digital for facile computer manipulation. It must be capable of assessing all the chemistry in the vast combinatorial field of possible routes, and then generating in abstract form the best routes from the field, only later refining the few best back to full chemical detail. Therefore, the nature of this generalizing system is critical to the success of the whole synthesis design process.

Characterization of Structures. The generalization used here (4,6,7) begins by defining just four synthetically important kinds of bonds which any single skeletal carbon may have. The σ-bonds to other carbons constitute the skeleton, but the π-bonds to other carbons are functional groups since they may be broken without altering the skeleton. The other bonds to carbon may be divided into bonds to more electronegative and to more electropositive atoms, since their balance determines the oxidation state and hence its change in a reaction.

These four kinds of bonds to any carbon may be labeled as R, Π, Z, H, respectively, as summarized in Figure 3, and the number of each kind on any carbon, i.e., σ, π, z, h, will add up to four. Any carbon can now be described with three variables: σ for its skeletal level, and $z\pi$ for its functional nature, with h as the difference of (σ+z+π) from 4. These three digits, $\sigma z\pi$, are a simple number which defines the *character* of any carbon. In this generalization, only 24 characters are possible, a number small enough to allow facile coverage of the great variety of organic structures while detailed enough to retain chemical significance (7).

The values of σ are the familiar descriptors of skeletal carbons: primary (σ=1), secondary (σ=2), tertiary (σ=3), and quaternary (σ=4). Only three values of π are possible for carbon atoms (π=0,1,2), distinguishing their attached single, double and triple bonds to other carbons, respectively.

Values of z (0-4) coalesce the common electronegative heteroatom functional groups as families of groups which are interchangeable among themselves by substitutions of one kind of heteroatom for another. Thus z=3 merges all derivatives of the carboxylic acid/nitrile family, and z=2 all derivatives of aldehydes and ketones (themselves distinguished by σ = 1 or 2, respectively). The z-values do not distinguish σ-bonds and π-bonds to the heteroatoms. Values of h (0-4) refer usually to hydrogens but also include the other electropositive atoms, or indeed a simple unshared electron pair on carbon, understood as the conjugate base of a bond to hydrogen.

An important product of this abstraction is that the oxidation state of any carbon in a structure is given by x = z - h, with values of x ranging from -4 to +4.

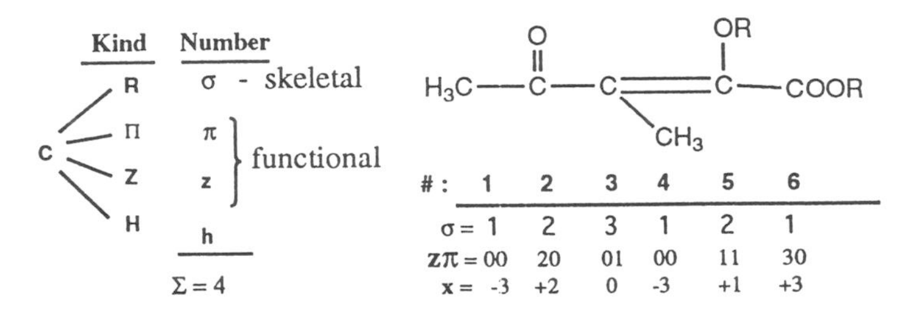

Oxidation State: $x = z - h$

16 Possible Unit Exchanges at a Skeletal Carbon

		$\Delta\sigma$	$\Delta\pi$	Δx
Substitution	HH, ZZ, RR, ΠΠ	0	0	0
Oxidation	ZH	0	0	+2
Reduction	HZ	0	0	-2
Elimination	ΠH	0	+1	+1
	ΠZ	0	+1	-1
Addition	HΠ	0	-1	-1
	ZΠ	0	-1	+1
Construction	RH	+1	0	+1
	RZ	+1	0	-1
	RΠ	+1	-1	0
Fragmentation	HR	-1	0	-1
	ZR	-1	0	+1
	ΠR	-1	+1	0

Figure 3. Generalized Characterization of Compounds and Reactions

Therefore, the oxidation state change in a reaction will be $\Sigma\Delta x$, over those carbons which change.

Characterization of Reactions. If structures can be generalized with simple character numbers ($\sigma z \pi$) for each carbon, then reactions can be described as the change in those numbers on passing from substrate to product. This constitutes the *net structural change* which characterizes any reaction family. The simplest reaction change is a single exchange of one bond for another on one carbon, i.e., one bond made and one bond broken. On any one carbon there are 16 possible single exchanges which are derived from the four generalized bond types. We label these changes at any carbon with a simple notation of two letters, the first for the bond made, the second for the bond broken, as shown at the bottom of Figure 3. These may be organized into the familiar reaction types: substitution, elimination/addition ($\Delta\pi$), and construction/fragmentation ($\Delta\sigma$), each with an oxidative and reductive variant (Δx). The three variables, $\Delta\sigma$, $\Delta\pi$, Δx, define the reaction change at each carbon (Figure 3) just as the number $\sigma z \pi$ defines its character.

The ideal synthesis is focused on construction reactions. A single construction reaction may be viewed at the C-C bond which is being made, with $\Delta\sigma$=1 for each joining carbon. The functional group changes which occur during construction take place on each carbon being joined and often on a further strand of carbons out from each one being joined as well. In Figure 4 these two strands of carbons are shown and labeled α,β,γ, out from each of the joining (α -) carbons. All the possible functional group changes that each of these strand carbons may exhibit are taken from Figure 3 and written out below the $\alpha\beta\gamma$ carbons.

The overall construction reaction may usefully be divided into two *half-reactions,* essentially independent of each other. There are three kinds of half-reactions listed in Figure 4 in terms of their strand length and structural change: simple substitution; addition; and allylic substitution, on strands of 1, 2, and 3 carbons, respectively. Most constructions are isohypsic, i.e., neither oxidative nor reductive overall, but their constituent half-reactions are all either oxidative or reductive. By adding Δx for each changing carbon from the table in Figure 3, the nucleophilic half-reaction changes in Figure 4 are seen to be oxidative, with $\Sigma\Delta x$ = +1, and the electrophilic half-reaction changes are reductive, with $\Sigma\Delta x$ = -1, so that joining one half-reaction of each general kind (3 x 3 = 9 ways) yields an overall isohypsic construction reaction.

Reactions which have only one single exchange of bonds at each changing carbon (the 16 exchanges of Figure 3) are *unit reactions*. In our survey of reaction databases (8), some 80% of all reactions are simply unit reactions and most of the rest are just composites of two sequential unit reactions, as with ZH+ZH to describe reduction of carboxyl to primary alcohol. Some common and useful constructions, like the Wittig reaction, are composites of a construction and a refunctionalization.

All of these reactions can be sharply defined with this system of description (9,10). In one class are the refunctionalizations with $\Sigma\Delta\sigma$=0; the rest are skeletal alterations, cf., constructions with $\Sigma\Delta\sigma$=2. These classes can be subclassified by

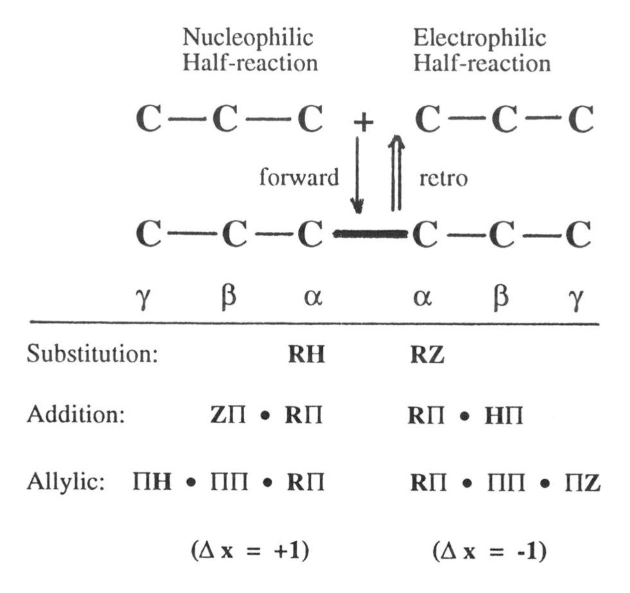

Figure 4. The Nature of Construction Reactions

the functional group change at each changing carbon, in the digital form of a "$\Delta z\pi$-list", the change in the z- and π-values at each changing carbon. Thus each reaction family has a numerical label, the $\Delta z\pi$-list, and they may be hierarchically subdivided further by a numerical list of the starting characters ($\sigma z\pi$) of the changing carbons.

This simple description of a reaction affords a facile procedure for the computer to generate products from substrates, or substrates from products, by adding the appropriate $\Delta z\pi$-list to the $z\pi$-values of each reacting strand of carbons - in the case of constructions, for the two half-reactions out from each end of the designated C-C bond.

Overview of the Route Generator

The whole procedure for generating synthesis routes will now have two phases: first, a skeletal dissection; and second, a functional group generation. In the first phase the target structure is stripped of its functional groups, and its skeleton is dissected into two intermediate skeletons in all possible ways. These are then cut in two again to afford four starting skeletons, and each set is accepted only if all four skeletons are found in the catalog of starting materials. Each such set defines an ordered bondset of the target skeletal bonds to construct, hence a general synthesis plan.

In the second phase, each of these ordered bondsets is taken separately, to generate the successive construction reactions needed to make its bonds. Starting with the known functional groups of the target as well as the last designated skeletal bond to make, it is now a simple matter to add all $\Delta z\pi$-lists to the $z\pi$-values of the carbons out each $\alpha\beta\gamma$ -strand from the joining carbons. For each successful generation, this gives a set of intermediate functional groups ($z\pi$) on the skeleton of the target minus one last skeletal bond, and so generates an intermediate compound.

This generation procedure is continued through the successive bonds of the ordered bondset until, on completing it, we arrive at sets of starting skeletons, now with $z\pi$-functional groups in place, defining real but generalized starting material molecules. These are compared with the catalog to establish whether they are represented by real compounds. Each of such generations, via $z\pi$-changes at affected carbons, constitutes a synthesis route and the only ones kept will be those which start with all four functionally defined starting materials found in the catalog.

The whole protocol is illustrated in Figure 5 for the synthesis of an estrone derivative. At left is one of the skeletal dissections, the same as in Figure 2. The target is cut into two intermediate skeletons (A and B) at the first level, and each is cut again in two at the second level; of course many other cut sequences are also found. This dissection is now retained because all four pieces (C, D, E, F) are found in the catalog; in fact, at the first level, B is also a found skeleton, giving this dissection some priority.

The second phase is shown starting from the target and passing down the right column of Figure 5. For each cut bond in the bondset, taken in order, is

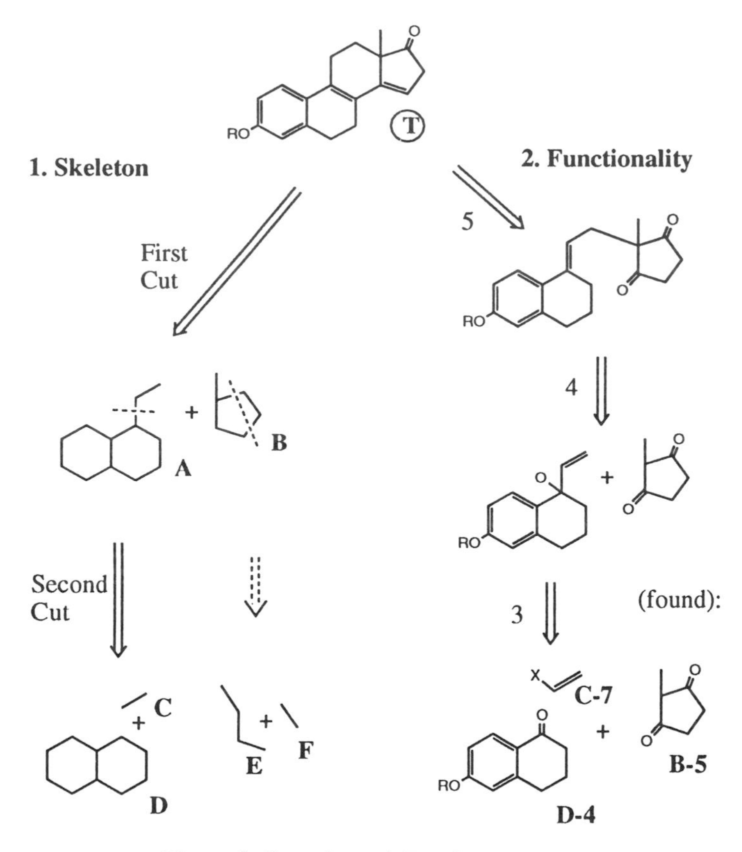

Figure 5. Overview of SYNGEN Protocol

shown one of the possible generated constructions consistent with the functional groups present. Others of course will also be generated. The zπ-values on the carbons are here pictured in common structure notation. Arriving at a full structure for skeleton B, the procedure finds that the intermediate generated is in fact a starting material, methyl-cyclopentanedione (B-5), and therefore does not need further cutting into E + F. The process must go further, however, to find the construction of intermediate A at the second cut level, which then generates the available starting materials C-7 and D-4, found among the various catalog compounds of those skeletons. The assembly plan for the generated bondset illustrated in Figure 5 is the same as that in Figure 2 except that forging bonds 1 and 2 in Figure 2 was not necessary since compound B-5 was found in the catalog. Furthermore, the actual generation of successive constructions has produced an ideal synthesis for this target molecule, itself two refunctionalization steps from estrone.

The SYNGEN Program

The protocol outlined above and illustrated in Figure 5 will produce a lot of ideal synthesis routes, in fact all of the best and shortest ones possible from an included catalog of starting materials. This system is implemented in the program SYNGEN with a catalog of some 6000 starting materials; many more can be added. In utilizing the procedure outlined above, this program is well suited to searching the synthesis tree for alternative synthesis routes, from which one may select those which are environmentally more benign.

The SYNGEN program is presently available in a number of academic and industrial locations as well as at the Environmental Protection Agency, and is generally found to be easy and user-friendly for the users in those institutions. The operation of SYNGEN is completely graphical. The target molecule is drawn on the screen with a fast, fluid drawing program, and the program takes the input target into batch mode to generate a set of syntheses, usually in under 5 minutes, allowing the user in this time either to enter another target to follow it in the batch, or to examine the output of previously generated syntheses deposited in a directory.

Validation of Generated Reactions. A major problem for the generator has always been that of recognizing and deleting unacceptable, non-viable reactions. The chemistry of viable reactions is of course more complex and variable than addressed by the simple, abstracted Δzπ-lists. In the past, reactions which were clearly not viable were deleted by a set of broad mechanistic tests, but many still slipped past this net and their presence tended to lower the credibility of the program in the eyes of its users. On the other hand, however, if the mechanistic tests are too restrictive, they will probably delete some viable routes. Hence another way to assess the quality of generated reactions was needed.

Now that large libraries of reactions are available in databases, it is possible for the SYNGEN program to validate the reactions it generates against this literature. To this end the COGNOS program has been developed as a reaction retrieval system to organize any reaction database for rapid searching (10). The COGNOS system is based on the same abstraction of the net structural change in a reaction, i.e., $\Delta z\pi$-lists, as that used in SYNGEN to generate new reactions. This makes it possible for SYNGEN to access the COGNOS index very rapidly, to find out the number of closely matching literature precedents and their average yield for each newly generated reaction. This index currently has over half a million literature reactions.

The routes generated may now be ordered within each bondset subtree by overall cost. This is calculated during the generation process from the starting material costs and the average yields per step from the literature precedents. Added to this is a cost for each reaction step, implying costs of solvents and reagents and the time it takes to execute a step. Thus the number of steps is strongly reflected in the overall cost.

The output from SYNGEN may be examined graphically in a number of ways which also allow for deletion by the user of unwanted reactions, intermediates, starting materials, or even overall bondsets. Also, a matrix of the number of routes by cost and number of steps may be accessed to focus examination on only a subset of them. Whole routes are presented, each complete on a whole screen with a description of the reaction types, the catalog starting materials, and the overall cost and steps. These may be examined one at a time in an order of preference based on cost or other criteria. When any whole route is displayed, unwanted compounds or reactions in it may be deleted directly on screen, so that they will not appear again in any routes displayed later.

It is now possible to add criteria of environmental acceptability to the generated synthesis routes. These will include quantitative measures such as acute and chronic toxicity, carcinogenicity, persistence, etc., for the starting materials in the catalog. These are being derived from lists of compounds available from the Environmental Protection Agency and linked directly into the route generation process in SYNGEN. From these lists it may be possible to generalize environmental concerns about the generated intermediates as well. In this way a second basis for ordering the generated routes may be superimposed on the costs so that a composite evaluation of routes may be used. The solutions can thus be ordered to bring to the fore those utilizing a minimum of toxic or hazardous chemicals en route.

Conclusions

A particular logic has been developed here for the design of organic syntheses. It is implemented in the SYNGEN program, which generates all the shortest synthesis routes to any given input target molecule, using as its source a catalog of available starting materials. The ease of operation of SYNGEN and its graphical clarity make it an ideal tool for teaching purposes, either for the practice of students in an academic setting or for the use of chemists in the chemical industry. An important goal of this program is to sensitize chemists and chemistry students to the great variety of possible and viable alternatives for the synthesis of target molecules.

This variety in turn provides a rich base for highlighting the differences in the environmental impact of these many alternate routes. The SYNGEN program will incorporate several measures of environmental hazards to emphasize these differences and to prioritize synthesis routes accordingly. Thus the SYNGEN program and its environmental assessments will make it possible to locate many possible more benign synthesis routes for compounds currently made by other paths.

This would make the consideration of environmental concerns systematic to the synthetic chemist *in the design stages first*, rather than having to deal later with the resultant problems of waste treatment, handling and disposal, as well as liability and workplace safety concerns.

Acknowledgment

We are grateful to the National Science Foundation for supporting the development of the SYNGEN program over the last decade, most recently under grant CHE-9403893.

Literature Cited

1. Corey, E. J.; Wipke, W. T. *Science* **1969,** *166,* 178.
2. Bersohn, M. *Bull. Chem. Soc. Japan* **1972,** *45,* 1897.
3. Chanon, M.; Barone, R. "Computer Aids in Chemistry", ed. Vernin, G.; Chanon, M., Ellis Horwood (1986), chap. 1.
4. Hendrickson, J. B. *Accts. Chem. Res.* **1986,***19,* 274; *Angew. Chem. Intl. Ed.* **1990,** *29,* 1286.
5. Hendrickson, J. B. *J. Am. Chem. Soc.* **1977,** *99,* 5439.
6. Hendrickson, J. B. *J. Am. Chem. Soc.* **1971,** *93,* 6847.
7. Hendrickson, J. B. *J. Chem. Ed.* **1978,** *55,* 216.
8. Hendrickson, J. B.; Miller, T. M. *J. Chem. Inf. & Comp. Sci.* **1990,** *30,* 403; *J. Am. Chem. Soc.* **1991,** *113,* 902.
9. Hendrickson, J. B. *Rec. trav. chim. Pays-bas* **1992,** *111,* 323.
10. Hendrickson, J. B.; Sander, T. L. *J. Chem. Inf. & Comp. Sci.* **1995**, *35*, 251.

RECEIVED September 21, 1995

Chapter 17

Incorporating Environmental Issues into the Inorganic Curriculum

James E. Huheey[1]

Department of Chemistry, University of Maryland, College Park, MD 20742

The addition of environmental topics, including green chemistry, to the traditional undergraduate inorganic course has been slow and uneven. However, this need not be the case. There are several examples of environmentally important problems that could be included in an inorganic chemistry course. These include problems that were anticipated and prevented, ones that were treated after the fact, and some that have not yet been solved satisfactorily. An argument is made for the inclusion of environmental chemistry with other descriptive chemistry in the sophomore "descriptive inorganic chemistry" course.

Despite the fact that it has been quite some time since Rachel Carson raised our consciousness with respect to the environmental impact of chemical pollution, courses in inorganic chemistry today do not contain much that is relevant to environmental chemistry. To be sure, general chemistry courses and some general chemistry textbooks may have significant amounts, but the inorganic chemistry courses per se, either with or without a physical prerequisite, generally contain few examples of environmental inorganic chemistry. This can be blamed for the most part on the tremendous amount of material vying for quite limited space in the usual one-semester course. Environmental inorganic chemistry is part of the larger body of descriptive inorganic chemistry, itself a matter of some contention (see below).

A survey of inorganic textbooks from the 1970's to the present with regard to topics on the environment and life science is shown in Fig. 1. The "good news" is that there has been a steady progression in the number of topics and the coverage of environmental issues. The "bad news" is that the coverage is quite uneven. There are

[1]Current address: Sourwood Mountain Scientific, 215 Tucker Lane, Lenoir City, TN 37771–3405

0097–6156/96/0626–0232$12.00/0

books published in the 1990's with no coverage at all. There certainly is a plethora of environmental topics that can be included in the inorganic chemistry course. These include examples both of the positive benefits of green chemistry and the negative effects of its absence. A few examples are discussed here.

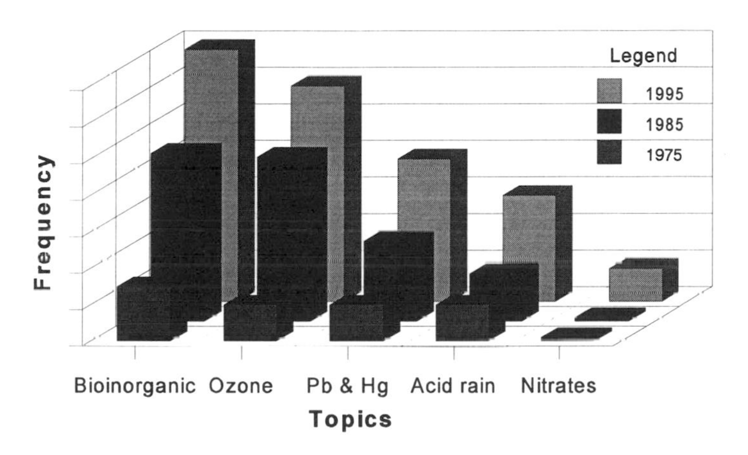

Figure 1. The frequency of selected topics of environmental and biological interest in inorganic textbooks in 1975, 1985, and 1995.

Descriptive Inorganic Chemistry in the Curriculum

There are a number of examples that could be used to illustrate green chemistry in inorganic chemistry courses. The fact that they have not been widely used is intimately tied to the problem that inorganic chemists have had confronting the issue of descriptive inorganic chemistry, of which environmental examples are a part. For some time, there has been a request, even a demand, for more descriptive inorganic chemistry in the undergraduate curriculum. Inorganic chemists have responded with a sophomore descriptive chemistry course that many schools offer. The American Chemical Society Committee on Professional Development has followed this lead and required one semester of (presumably largely descriptive) inorganic chemistry in addition to the traditional advanced course with a physical chemistry prerequisite. However, the option of having either a separate course or the incorporation of the material into the second semester of the first-year general chemistry course has proved to be too much of a temptation to most curriculum planners, pressed for maximum material and restricted by the very real constraints of time. Thus many schools do not offer such a course (*1*).

A second problem facing those attempting to get more descriptive inorganic chemistry (including environmental aspects) through the adoption of a standard sophomore descriptive course has been the lack of textbooks for such a course. This has tended to be a circular problem because there has not been much enthusiasm in writing a book for a very small market.

In summary, although there is not much environmental chemistry in inorganic courses, this is not for lack of material. There are abundant examples available for use. It may be hoped that, like recently introduced material on bioinorganic chemistry, more environmental chemistry will work its way into these courses.

Specific Problems with Inorganic Pollutants

Traditionally, inorganic chemistry has been a great contributor to our environmental problems. Historically problems of this type arose long before the current environmental crisis. Most often inorganic pollution results from the presence and concentration of a specific element rather than from the particular molecule in which the element is found in nature. Mercury poisoning and *itai itai disease*, lead and its multiple toxic effects, or any of the problems associated with high concentrations of certain transition metals all relate to the specific element.

Consider how this situation differs from that of an organic pollutant. The latter is ultimately only a collection of atoms of carbon and hydrogen, and perhaps oxygen, nitrogen, sulfur, or a halogen. Its hazard and subsequent polluting potential comes from the way those atoms are arranged. If the arrangement is changed, the properties are changed. In principle, the molecule can be converted into another molecular form or product that might be useful. It may be biodegradable. If nothing better is available, it can probably be incinerated. This is not to imply that organic pollutants are no problem, merely that they tend to be qualitatively different from elemental inorganic pollutants such as lead, mercury, etc. In the latter case the nature of the molecular form may affect the solubility, distribution, fractionation, availability, even the toxicity of the element, but ultimately it is less important than the presence of the toxic element itself.

The fundamental hazard of inorganic chemicals comes from the fact that these elements would not normally be a part of the ecosystem. So as they are transformed, whether by synthetic processes, biological cycles, or by incineration and broadcasting over the face of the earth, in some ways they remain unchanged--elements are indestructible. And, as is well known, dilution is not a solution to pollution, since these toxic elements can be concentrated biologically until they again pose a problem. So the problems arising from toxic elements come from the use in manufacturing processes of elements that are not normally present in biological systems.

Table I lists a few elements with a rough classification biologically as "Essential", "Neutral", or "Toxic", and a rough classification with respect to abundance in sea water. The criteria for abundance are arbitrary: anything as common as iron is considered "Abundant" and includes the common metals and nonmetals used biologically. "Rare" consists of the less common transition metals that are still sufficiently abundant to be used in enzyme systems. "Very rare" are metals that are quite rare (e.g. precious metals and other heavy metals), or more specifically, 1) they are so rare that they were unavailable for life to use as it was evolving various mechanisms and processes, and 2) because they were so rare, life not only did not use them, it also did not develop mechanisms to handle them in an innocuous, nontoxic manner. Upon the introduction of these elements into the environment through manufacturing processes, living systems were subjected to these substances and could not handle them adequately.

Table I. Natural abundance of some elements related to their Biological Properties. Abundance in Sea Water *(2)*

	Abundance in Sea Water		
Biological Properties	Common ≥ 0.003 mg L^{-1}	Rare $\sim 10^{-3}$ mg L^{-1}	Very Rare $\sim 10^{-6}$ mg L^{-1}
Essential	H, B, C, N, O, F, Na, Mg, Fe, Cu, Zn, Mo	V, Cr, Mn, Ni, Co	
Neutral	Li, Rb	Ti	
Toxic			Be, Au, Hg, Tl, Pb

Copper Basin, Tennessee *(3-5)*

The main chain of the Appalachian Mountains, separating the Gulf of Mexico from Atlantic drainage defines the state line between North Carolina and Tennessee. This area is mostly wooded, especially along the backbone of the southern Appalachian Mountains. Just west of the North Carolina border and slightly north of the Georgia border lies Ducktown, TN. This area is known as the Copper Basin. It consists of fifty square miles of deforested land in the midst of the main Appalachian chain of forests. Until recently, it was the third anthropogenic artifact, together with the Great Wall of China and the Great Pyramids, that was visible to the naked eye from space.

Copper was discovered here in 1843, and the first smelter began operating in 1850. The Copper Basin mines and smelters produced 90% of the copper used by the

Confederacy in the Civil War until northern troops occupied the area in 1863. The copper ore was first roasted in open fires consisting of alternate layers of wood and ore. These ricks of wood and ore were several feet high and up to 200 feet long, and burned over a period of two to three months. This preliminary roast was necessary because of the pyritic nature of the ore. Both the roasting and the later smelting converted the sulfur to sulfur dioxide. This was allowed to escape into the air where it was oxidized into sulfuric acid providing the first large scale example of acid rain in this country. With a molecular weight of 64, more than twice the average molecular weight of air, the SO_2 settled in the basin. Before reforestation was undertaken, it was possible to trace the level to which the SO_2 had risen - there was a deforested contour around the mountains. In the old days the atmosphere was so thick that the work mules all wore bells to keep from bumping into each other. Still, it wasn't lethal - that would have been self-limiting.

Between the cutting of timber and subsequent acid conditions, more than 130 km^2 (50 sq. miles) were completely defoliated (*6*). Around the turn of the century processes were developed that allowed the SO_2 to be captured and converted to sulfuric acid. This not only put an end to first example of acid rain in this country, it also proved to be a profitable endeavor that continued even after the mines were closed and sulfur had to be shipped in to Ducktown for the manufacture of sulfuric acid! This is just one case that demonstrates well the advantages of benign synthesis. If it had been possible to obtain both copper and sulfuric acid initially, the damage could have been averted.

Homogeneous Catalysis

Homogeneous catalysis is one of the success stories of modern inorganic chemistry (see Figure 2 for an example). As George Parshall has pointed out, the organic chemical industry has progressed from the days when the basic building block was acetylene which was followed later by the less reactive alkenes. Today we are progressing to the age of alkanes and synthesis gas (*7*). Homogeneous catalysis has made this progress possible providing numerous advantages such as lower temperatures of reactions energy savings. Also the lower temperatures tend to provide greater specificity and fewer unwanted, and perhaps undesirable, by-products. Thus ideally these catalysts lead the way towards a totally benign synthesis: minimum energy costs coupled with maximum yields and purity.

There is one major disadvantage, however: the need to separate and recover the catalyst. This disadvantage has two aspects. Cost aside, the release into the biosphere of large amounts of the heavier transition metals would almost certainly have untoward effects. They are the very elements discussed in an earlier section. They are rare and have never occurred in appreciable amounts in living systems. But here, too, there is an encouraging lesson to be learned - the catalysts are compounds of very expensive metals - any significant loss would be prohibitively expensive. This very fact has forced the development of methods to prevent their loss. It is mainly a matter of the necessary incentive.

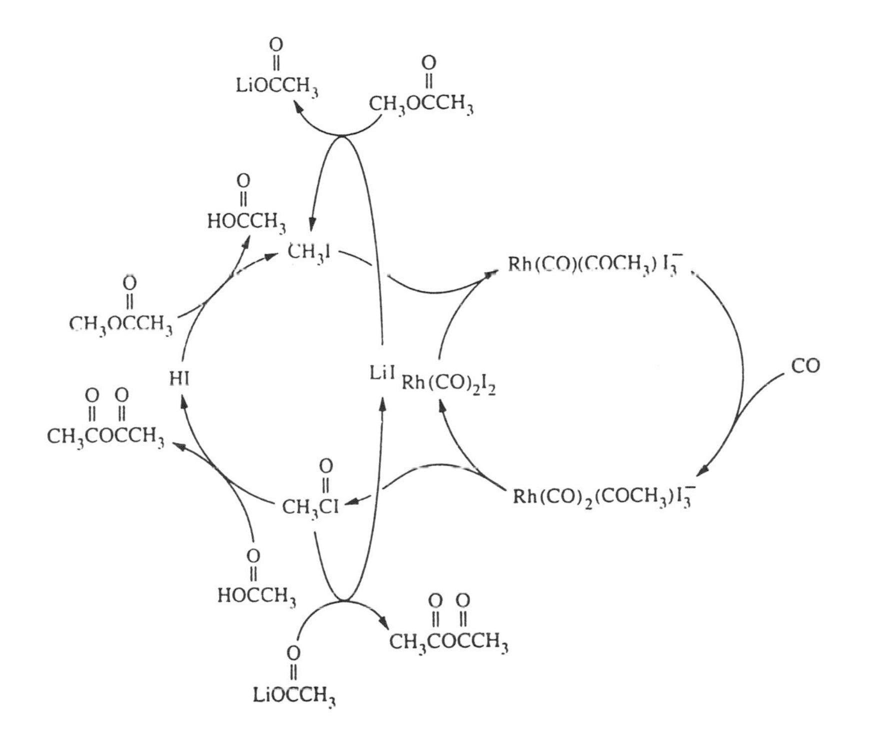

Figure 2. A typical homogeneous catalytic cycle - the Tennessee Eastman process for the conversion of methyl acetate to acetic anhydride via carbon monoxide insertion. [From reference 2, page 713. Reproduced with permission.]

Conclusions

There is certainly much environmental chemistry that could be included in the inorganic curriculum. Although much inorganic environmental chemistry is related to toxic elements as described early in this chapter, one specific example of an appropriate molecular reaction with environmental relevance that could be incorporated into inorganic chemistry courses is the destruction of the ozone layer. This reaction involves both photochemistry and catalysis by chlorine or nitrogen oxides:

$$F_3CCl + h\nu \ (190\text{-}220\ nm) \rightarrow F_3C^{\cdot} + Cl^{\cdot}$$

$$Cl + O_3 \rightarrow ClO + O_2$$

$$ClO + O \rightarrow Cl + O_2 .$$

The net reaction is:

$$O + O_3 \rightarrow 2O_2 .$$

Other examples of environmental chemistry that could be incorporated into inorganic chemistry courses include both the prevention of pollution as heavy metals are used or specific benign syntheses developed for industrial processes.

The future can only increase the quantity and quality of examples available. Environmental chemistry is a legitimate part of descriptive inorganic chemistry and deserves a place along with descriptive chemistry either in a second semester of general inorganic chemistry or as a separate course.

Literature Cited

1. The author can speak from direct experience with second semester general chemistry courses that are inorganic only on the forms submitted to the ACS. Fortunately, in our department we offer *two* upper level courses, not requisites, that are taken by a large number of our students. One course treats the descriptive chemistry of the transition metals, the other that of the main group elements.
2. Huheey, J. E.; Keiter, E. A.; Keiter, R. L. *Inorganic Chemistry: Principles of Structure and Reactivity, Fourth Edtion;* HarperCollins Publishers: New York, **1993**.
3. Quinn, M.-L. *J. Soil Water Cons.* **1988**, *43*, pp. 140-144.
4. Quinn, M.-L. Atmos. Environ. **1989**, *23*, pp. 1281-1292.
5. *Quinn, M.-L. Environ.Management,* **1991**, *15*, pp. 179-194.
6. Emmons, W. H.; Laney, F. B. *U. S. Geol. Surv. Pap.13R,* U. S. Government Printing Office: Washington, DC, **1926**.
7. Parshall, G. W. *Homogeneous Catalysis,* Wiley: New York, **1980**, p. *223*.

RECEIVED December 29, 1995

INDEXES

Author Index

Affiliation Index

Subject Index

L

M

N

O

P

Production: Charlotte McNaughton & Amie Jackowski
Indexing: Zeki Erim, Jr.
Acquisition: Rhonda Bitterli & Barbara E. Pralle
Cover design: Paul T. Anastas, Tracy C. Williamson, & Michelle Telschow

Printed and bound by Maple Press, York, PA